Lecture Notes of the Institute for Computer Sciences, Social Informatics and Telecommunications Engineering

403

More information about this series at https://link.springer.com/bookseries/8197

Tadashi Nakano (Ed.)

Bio-Inspired Information and Communications Technologies

13th EAI International Conference, BICT 2021
Virtual Event, September 1–2, 2021
Proceedings

 Springer

Editor
Tadashi Nakano 📵
Osaka City University
Sugimoto, Japan

ISSN 1867-8211   ISSN 1867-822X (electronic)
Lecture Notes of the Institute for Computer Sciences, Social Informatics
and Telecommunications Engineering
ISBN 978-3-030-92162-0   ISBN 978-3-030-92163-7 (eBook)
https://doi.org/10.1007/978-3-030-92163-7

This Springer imprint is published by the registered company Springer Nature Switzerland AG
The registered company address is: Gewerbestrasse 11, 6330 Cham, Switzerland

Preface

We are delighted to introduce the proceedings of the 13th EAI International Conference on Bio-inspired Information and Communications Technologies (BICT 2021). Looking back, the first edition of this conference was held 15 years ago as BIONETICS (the International Conference on Bio-inspired Models of Network, Information and Computing Systems) in 2006 with the goal of facilitating research and development of bio-inspired ICT. The following editions of the conference since then have been held almost annually, providing a world-leading venue for researchers and practitioners to discuss recent results on bio-inspired ICT. Consistent with the goal of prior editions, EAI BICT 2021 was organized. Due to the safety concerns and travel restrictions caused by COVID-19, however, BICT 2021 took place online as a virtual interactive conference September 1–2, 2021.

The technical program of BICT 2021 consisted of two keynote talks and 21 paper presentations. On day 1, Takahiro Hara of Osaka University, Japan delivered his keynote talk on "Cross-domain User Activity Prediction: Todays and Future"; his recent work on cross-domain user activity prediction was presented, including discussion on how bio-inspired approaches might be integrated to deal with unknown or uncertain situations. Day 1 also included two technical sessions: "Bio-inspired Network Systems and Applications" (8 paper presentations) and "Bio-inspired Information and Communication" (7 paper presentations). Day 2 began with another keynote by Dan V. Nicolau of McGill University, Canada. Entitled "Biocomputation with Motile Biological Agents", he presented his recent research results of using microorganisms for solving computational problems. The conference then proceeded to the last technical session on "Mathematical Modelling and Simulations of Biological Systems" (6 paper presentations) followed by the closing remarks.

This year, we received 47 paper submissions, and accepted 22 papers, including 20 full papers and 2 short papers. We appreciate our Program Committee members for their hard work in reviewing papers carefully and rigorously. With our congratulations to the authors of accepted papers, the BICT 2021 conference proceedings consists of 21 high-quality papers (one paper withdrawn).

The organization of the BICT 2021 conference proceedings relies on the contributions by Organizing Committee members as well as PC members. It was our privilege to work with these respected colleagues. Last but not least, special thanks go to the EAI, particularly Karolina Marcinova, for helping us organize BICT 2021 and publish this proceedings successfully.

Tadashi Nakano
Andrew Eckford
Yifan Chen
Kun Yang
Sasitharan Balasubramaniam
Yutaka Okaie

Organization

Steering Committee

Imrich Chlamtac	University of Trento, Italy
Jun Suzuki	University of Massachusetts, Boston, USA
Tadashi Nakano	Osaka City University, Japan

Organizing Committee

General Chair

Tadashi Nakano	Osaka City University, Japan

TPC Chair and Co-chair

Andrew Eckford	York University, Canada
Yifan Chen	University of Electronic Science and Technology of China, China
Kun Yang	University of Essex, UK
Sasitharan Balasubramaniam	Waterford Institute of Technology, Ireland
Yutaka Okaie	Osaka University, Japan

Sponsorship and Exhibit Chair

Huber Nieto	Universidad Autonoma del Peru, Peru

Workshops Chairs

Lin Lin	Tongji University, China
Mohammad Upal Mahfuz	University of Wisconsin-Green Bay, USA

Publicity and Social Media Chair

William Casey	United States Naval Academy, USA

Publications Chair

Qiang Liu	University of Electronic Science and Technology of China, China

Web Chair

Peng He   Chongqing University of Posts and
                            Telecommunications, China

Technical Program Committee

Aftab Ahmad   John Jay College of Criminal Justice, USA
Hamidreza Arjmandi   Yazd University, Iran
Rafael Asorey-Cacheda   Universidad Politécnica, Spain
Hamdan Awan   Waterford Institute of Technology, Ireland
Xu Bao   Lanzhou University, China
Chan-Byoung Chae   Yonsei University, South Korea
Chun Tung Chou   University of New South Wales, Australia
Uche Chude-Okonkwo   University of Johannesburg, South Africa
Debanjan Das   International Institute of Information Technology,
                               India
Yansha Deng   King's College London, UK
Douglas Dow   Wentworth Institute of Technology, USA
Faramarz Fekri   Georgia Institute of Technology, USA
Luca Felicetti   University of Perugia, Italy
Mauro Femminella   University of Perugia, Italy
Hiroaki Fukuda   Shibaura Institute of Technology, Japan
Vinay Gautam   Aalto University, Finland
Krishnendu Ghosh   College of Charleston, USA
Preetam Ghosh   Virginia Commonwealth University, USA
Weisi Guo   University of Warwick, UK
Werner Haselmayr   Johannes Kepler University, Austria
Henry Hess   Columbia University, USA
Li Huang   Tongji University, Shanghai China
Adam Noel   University of Warwick, UK
Pratip Rana   U.S. Army Engineer Research and Development
                               Center, USA
Satyaki Roy   University of North Carolina, USA
Hiroyuki Sato   The University of Electro-Communications, Japan
Robert Schober   Friedrich-Alexander Universität, Germany
Fan-Hsun Tseng   National Taiwan Normal University, Taiwan
Chenggui Yao   Shaoxing University, China
Cevallos Yesenia   National University of Chimborazo, Italy

Contents

Mathematical Modelling and Simulations of Biological Systems

Bio-inspired Network Systems
and Applications

Multi-objective Optimization Deployment Algorithm for 5G Ultra-Dense Networks

Yun-Zhe Li[1], Wei-Che Chien[2], Han-Chieh Chao[3], and Hsin-Hung Cho[1(✉)]

[1] Department of Computer Science and Information Engineering, National Ilan University, Yilan, Taiwan
[2] Department of Computer Science and Information Engineering, National Dong Hwa University,
Hualien City, Taiwan
[3] Department of Electrical Engineering, National Dong Hwa University,
Hualien City, Taiwan

Abstract. Due to insufficient spectrum resources, B5G and 6G will adopt millimeter waves for data transmission. Due to the poor physical characteristics of millimeter-wave diffraction ability, a large number of base stations are required for deployment, forming ultra-dense networks. Regarding the deployment of base stations, the first problem faced by operators is how to optimize the deployment of base stations in consideration of deployment costs, coverage rates and other factors. This research focuses on multi-objective three-dimensional (3D) small cell deployment optimization for B5G mobile communication networks (B5G). An optimized deployment mechanism based on NSGA-II is proposed. The simulation results show that, compared with NSGA-II, the deployment cost of this method is slightly higher, but it has achieved better results in terms of coverage and RSSI indicators.

Keywords: Multi-objective optimization · 3D base station deployment · NSGA-II

1 Introduction

With the rapid development of wireless communication, the frequency spectrum resources within 30 GHz are almost exhausted. To meet the requirements of 5G, it is necessary to solve the problem of insufficient spectrum. Therefore, the 5G New Radio (NR) [1] has determined to mainly use two frequencies, including the Frequency Range 1 (FR 1) band and the FR 2 band. The frequency range of the FR1 band is from 450 MHz to 6 GHz, also called the sub 6 GHz band; the frequency range of the FR 2 band is from 24.25 GHz to 52.6 GHz, also called mmWave [2,3]. The atmosphere will selectively absorb electromagnetic waves of

© ICST Institute for Computer Sciences, Social Informatics and Telecommunications Engineering 2021
Published by Springer Nature Switzerland AG 2021. All Rights Reserved
T. Nakano (Ed.): BICT 2021, LNICST 403, pp. 3–14, 2021.
https://doi.org/10.1007/978-3-030-92163-7_1

certain frequencies (wavelengths) in radio waves propagation, causing the path loss of these electromagnetic waves to be particularly serious. Electromagnetic waves are mainly absorbed by oxygen and water vapor. The resonance caused by water vapor absorbs electromagnetic waves between 22 GHz and 183 GHz, and the resonance absorption of oxygen affects electromagnetic waves between 60 GHz and 120 GHz.

Therefore, no matter which organization allocates mmWave resources, it needs to avoid the frequency bands around these four frequencies. In addition, mmWaves above 95 GHz are not considered for use due to technical difficulties. Although the use of mmWave can provide greater bandwidth and higher transmission rate, it has the problem of too large path loss and too small coverage [4], which makes the deployment of 5G base stations more difficult. In addition, the penetration of radio waves comes from the diffraction characteristics of radio waves. Generally speaking, the low-frequency signal has a diffraction effect because of its longer wavelength. The diffraction allows us to receive signals through the wall but mmWave signals have relatively low wall penetration due to high frequency, short wavelength, and short diffraction radius. For the above reasons, a large number of 5G base stations will be deployed and an Ultra-Dense Heterogeneous Network (UDHN) architecture will be formed [5–7]. In order to effectively reduce the deployment cost. This study formulated deployment problem of 5G cellular network as a multi-objective optimization problem, considering deployment cost, Received Signal Strength Indication (RSSI), coverage to achieve optimal deployment. Furthermore, we also considered the interference of obstacles and proposed a 3D deployment algorithm based on NSGA-II.

The rest of this dissertation is organized as follows. In Sect. 2, we introduce the current study of small cell deployment. Then, the problem definition is shown in Sect. 3. Section 4 presents improved NSGA-II algorithm for deployment problem. The results analysis is shown in Sect. 4.2. Section 5 provides conclusion.

2 Background and Related Works

For deployment problems, coverage is often used as an objective function [8–10] pointed out that the deployment density of 5G ultra-dense cellular networks will affect RSSI. The too sparse deployment will lead to coverage holes but too dense will increase interference between cells. [11] uses a single-objective genetic algorithm to deploy wireless sensor nodes. The largest coverage area minus the overlapping area of coverage is taken as the objective function. The single-target problem is difficult to optimize the trade-off relationship between different parameters. Therefore, [12] proposed a multi-target deployment problem, considering the ratio of deployment cost, signal, interference, and coverage. Then, they choose the best deployment plan from the set of candidate positions. [13] proposed a method for optimizing base station deployment based on deployment costs. They consider three network nodes, including Macro cell, Pico cell, and relay node, and select the best location from the candidate locations. In [14], the difference from the above method is that there is no candidate location design. The deployment location can be arbitrarily selected, providing a greater degree of freedom and making the location of the base station closer to the best solution.

3   Problem Definition

The defined planning problem of 5G small cell deployment is called a Signal Quality Maximization (SQM) problem. The SQM problem aims to maximize the signal quality of each served user by avoiding interference from buildings. Then, the SQM problem has been reduced and mapped to a well-known NP-complete problem, such as the Knapsack problem. Therefore, the SQM problem is proved as an NP-complete problem [14].

Small cell deployment problems can be close to an actual environment by using the results of traffic prediction. The scenario of this dissertation is in the 5G wireless backhaul network. There are K base stations, P users. Where B_k represents the kth base station and U_p represents the pth user. The wireless frequency of communication between B_k and U_p is represented by f. The topic of this dissertation is how to design a base station deployment scheme that meets the requirements of operators and users. Therefore, the coverage rate, RSSI, and deployment cost of the base station are considered. However, these three factors have a trade-off problem. In fact, when other factors remain unchanged, the number of base stations is reduced, which can intuitively reduce the deployment cost of the base station but it may also lead to a reduction in coverage or RSSI. Therefore, the location of the base station is relatively important because the same base station in different locations may cause different coverage. The base station at the appropriate location can reduce deployment costs. The evaluation function of the deployment strategy of the base station based on the above problems is defined in this dissertation.

Before evaluating the RSSI and coverage rate, the following values must be known the relative distance between base stations, the signal strength received by the indoor user, and the relative distance between the building and the base station. Figure 1 shows the distance relationship between the user and the building. If the user is inside the building, the distance between the outside and inside of the building needs to be additionally considered.

$$D_{k,p}^{3D} = d_{k,b}^{3D} + d_{b,p}^{3D} \tag{1}$$

$$D_{k,p}^{2D} = d_{k,b}^{2D} + d_{b,p}^{2D} \tag{2}$$

$$D_{k,p}^{3D} = \sqrt{D_{k,p}^{2D} + (H_k^{b,s} - H_p^{u,s})} \tag{3}$$

Where $d_{k,b}^{3D}$ is the distance between B_k and the outside of building in three-dimensional, $d_{b,p}^{3D}$ is the distance between U_p and the inside of building in three-dimensional, $d_{k,b}^{2D}$ is the distance between B_k and the outside of building in two-dimensional, $d_{b,p}^{2D}$ is The distance between U_p and the inside of building in two-dimensional, $H_k^{b,s}$ indicates the height of B_k, and $H_p^{u,s}$ is the height of U_p. The above units are all meters, used to describe the distance relationship between the base station, users, and buildings.

3.1   Received Signal Strength Indicator

RSSI is used to estimate the signal strength in a wireless network environment, affected by the interference of neighboring channels and obstacles. Neighboring channel interference means that when there are neighboring base stations with the neighboring channel, the generated radio waves will interfere with the user's signal. However, in current or future base station deployment planning, base stations with different frequencies will be arranged next to any base station as far as possible to avoid the above-mentioned neighboring channel interference. In addition, when a user enters an environment with a wireless network, there may be multiple base stations that can be connected. Usually, users will be connected to the base station which has the best reception. The RSSI received by the device can be calculated by Eq. (4).

$$f_{RSSI} = \frac{\sum_{p=1}^{N} R_p}{N}, R_p = \begin{cases} T_k - P_{k,p}, & \text{if } C_{k,p} = 1 \\ 0 & \text{otherwise} \end{cases}, \tag{4}$$

Where N is the number of UEs. R_p is received signal strength indicator, defined as the signal power at the transmitter minus the path loss. T_k is the Transmitter Signal Strength Indicator (TSSI) of B_k with dBm unit. $P_{k,p}$ is path loss between B_k and U_p with dBm unit. $C_{k,p}$ is a Boolean value and indicates whether U_p is covered by B_k as shown in Eq. (5).

$$C_{q,k} = \begin{cases} 1, & \text{if } D_{k,p}^{3D} < TH_D \text{ and } R_p > TH_R \\ 0, & \text{otherwise} \end{cases}, \tag{5}$$

In addition to considering the distance between the user and the base station, the RSSI actually received by the user also needs to larger than the threshold TH_R. Where $D_{k,p}^{3D}$ is the distance between B_k and U_p in three-dimensional, TH_D is a threshold, indicating the longest distance that the base station communicates with the user in LOS.

$$P_{k,p} = p_{k,p}^{\alpha} + p_{k,p}^{\beta} + p_{k,p}^{\gamma} \tag{6}$$

Where $P_{k,p}$ is path loss between B_k and U_p, $p_{k,p}^{\alpha}$ is basic path loss, $p_{k,p}^{\beta}$ is penetration loss, and $p_{k,p}^{\gamma}$ is indoor path loss.

3.2   Coverage Rate

Coverage rate is an important indicator for users. If an operator can provide a networked service anytime and anywhere, users will naturally choose this operator. Therefore, in addition to signal power, coverage is also the key for operators to attract users. Equation (7) is the calculation of coverage.

$$f_{coverage} = \frac{\sum_{p=1}^{N} C_v}{N}, C_v = \begin{cases} 1, & \text{if } \sum_{k=1}^{Q} C_{k,p} \neq 0 \\ 0, & \text{otherwise} \end{cases}, \quad (7)$$

Q represents the number of the base stations. $C_{k,p}$ is a Boolean value which means whether U_p is covered. In other words, it represents whether the user is covered by any base station. C_v represents the sum of covered U_p. If C_v is equal to 0, it means that U_p is not covered by any B_k. Finally, dividing all C_v by the total number of users P to get the coverage rate for users.

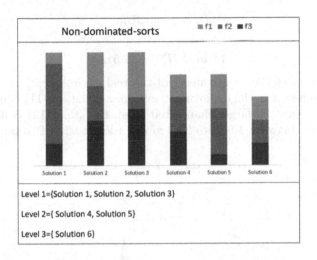

Fig. 1. The schematic diagram of Non-dominated-sort classification.

3.3 Deployment Cost

Deployment cost is very important for the operator. The deployment of the base station may result in the same cost but different coverage or RSSI due to the location. Therefore, in addition to coverage and RSSI, the deployment cost must be considered. The main deployment cost of the wireless network comes from the base station. We define the main cost as follows Eq. (8). Where f_{cost} is the objective function of deployment cost and Lis the set of deployment cost with$L = \{l^{MC}, l^{SC1}, l^{SC2}, l^{SC3}\}$. l^{MC} is deployment cost of Macro cell, l^{SC1} is deployment cost of type 1 small cell, l^{SC2} is deployment cost of type 2 small cell, and l^{SC3} is deployment cost of type 2 small cell. A, B, C, and D are the sum of cells. c_0, c_1, c_2, and c_3 are the number of cells.

$$f_{cost} = \sum_{c_0=1}^{A} l^{MC} + sum_{c_0=1}^{B} l^{SC1} + sum_{c_0=1}^{V} l^{SC2} + sum_{c_0=1}^{D} l^{SC3}, \quad (8)$$

$$A + B + C + D = K \quad (9)$$

The deployment of base stations needs to be considered from the perspective of users and operators to provide low RSSI, high coverage, and low deployment costs. Therefore, this dissertation formulates the deployment problem as a Multi-Objective Optimization (MOO) problem according to the above three factors, as shown in Eq. (10).

$$f(x) \begin{cases} \text{Max } f_{RSSI}(x) \\ \text{Max } f_{coverage}(x) \\ \text{Min } f_{cost}(x) \end{cases} \tag{10}$$

$$R_p > -100 \, \text{dBm} \tag{11}$$

$$1.5 \, \text{m} < H_p^{ue} < 22.5 \, \text{m} \tag{12}$$

In order to meet the requirements of the real environment, we define some restrictions. Where x is the deployment solution. Equation (11) is used to limit RSSI of U_k. It must be larger than -100 dBm. Equation (12) is used to limit the height of U_k between 1.5 m to 22.5 m according to 3GPP standard [15].

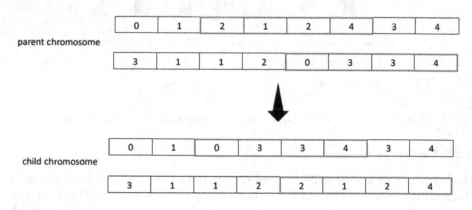

Fig. 2. Schematic diagram of crossover step.

3.4   Improved Non-dominated Sorting Genetic Algorithm II

The key to the multi-objective optimization problem is to find the best solution set of Pareto. Many algorithms have been proposed based on the concept of Pareto improvement, the most classic of which is the NSGA-II proposed by Kalyanmoy Deb and others in 2000 [16]. This algorithm is applied to multi-objective problems based on the genetic algorithm. The algorithm uses the concept of mating and mutational biomimetic evolution of the genetic algorithm, combined with the Non-dominated-sort method to reduce the complexity of the calculation.

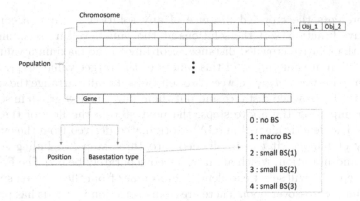

Fig. 3. Schematic diagram of chromosomes and genes.

Then NSGA-II adopts the elite strategy and merges the parent and offspring populations to ensure that excellent solutions can continue to participate in the algorithm and improve the optimization results in terms of results. In addition, NSGA-II introduces the calculation of crowding distance to compare solutions of the same Pareto level with large differences to ensure the diversity of the next generation and avoid similar results from mating with each other. The concept is like avoiding inbreeding and losing diversity.

In addition, NSGA-II also has the advantages of fast operation speed and good convergence performance. NSGA-II's Non-dominated- sort is to classify the solution set according to the function of each solution for each goal to obtain $s_l.s_l$ represents the set of solutions of level 1 after a Non- dominated-sort. This is a cyclical process of grading according to the target value. First, find the Non-dominated solution set in the group, which is counted as s_1, and remove it from the entire set S. Then continue to find the Non-dominated solution set in the next level group and record it as s_2. This process is repeated until all classifications are completed. Figure 1 is a schematic diagram of Non-dominated-sort classification.

The meaning of crowded distance is to find a solution with large differences. As the mixing of parent and children in NSGA-II will exceed the number of original chromosomes, after mixing the population, they need to be classified by Non-dominated-sort ($S = \{s_1, s_2, \ldots, s_l\}$). Then, according to the classification results, put s_l in order until the s of a certain level cannot be completely put into the new solution set. At this time, the solution to be put in priority should be selected according to the solution's Crowding distance. The calculation of crowding distance is shown in Eq. (13). The crowded distance of the solutions is after non- domination-sort, the objective function gap between the solutions j at the first level and the adjacent solutions $j + 1$ and $j - 1$.

$$w_s = \frac{m_{s+1} - m_{s-1}}{m_{max} - m_{min}}, \tag{13}$$

w_s represents the crowded distance of any solution s. First, according to the objective function, sort the solutions at the same level in ascending order. Then, get the relative crowded distance according to the maximum value m_{max} and the minimum value m_{min} of this function, the target value m_{s+1}, and the distance ratio between m_{s-1} between two solutions. Finally, arrange the solutions according to the crowded distance and put them into the new group in sequence.

After completing the above steps, the next step is the flow of the genetic algorithm. The genetic algorithm (GA) is originated derived from the evolutionary theory of biology. It can be divided into three steps, including selection, crossover, and mutation, which are introduced separately below. The first is the selection step. GA will select excellent chromosomes from the solution space and put them into a crossover pool. There are many selection methods has proposed. This dissertation uses the roulette method according to the objective function value of each chromosome that gives the corresponding selection probability. In NSGA-II, the roulette rate is based on the solution that belongs to which F_t. Therefore, the roulette rate is determined after the chromosome is selected into the crossover pool. Then, steps of crossover and mutation are performed in sequence according to the crossover rate and mutation rate. The crossover step is to randomly take two chromosomes for partial chromosome exchange, as shown in Fig. 2. The mutation step is to randomly select a gene from the chromosome to change.

The following describes the 3D small cell deployment algorithm proposed by this research. First, each chromosome represents a deployment solution (CS). Figure 3 is a schematic diagram of chromosomes and genes. Genes represent the type and position coordinates of each base station. There are two parameters in each gene, including position and type of the base station. The number 1 is a Macro cell, the numbers 2 to 4 are three different types of small cells, and the number 0 means there is no base station at this location. In addition, the location of the base station may be located in the building, at the top of the building, or on the telephone pole.

Therefore, when the gene is changed by steps of crossover or mutation, it means that the type or location of the corresponding base station in the deployment plan has changed. For the convenience of storage and calculation, the objective function is stored at the end of the chromosome.

However, the NSGA-II algorithm may cause too long computational time due to complicated calculations on the base station deployment problem, making the solution difficult to converge. Therefore, this dissertation uses the annealing mechanism of the simulated annealing algorithm to change the length of crossover and mutation. In the beginning, the algorithm has a strong search space. As the number of iterations increases, the search space in the latter part of the algorithm will shrink, prompting the results to gradually converge. The length of crossover and mutation is based on Eq. (14).

$$CL_m - CL_{m-1} \times (1 - \frac{m}{M}), \tag{14}$$

Where m indicates the mth iteration, M is the total number of iterations, CL_m is the length of the gm^{th} crossover. According to Eq. (14) as the iteration increases, the length of crossover will become shorter and shorter.

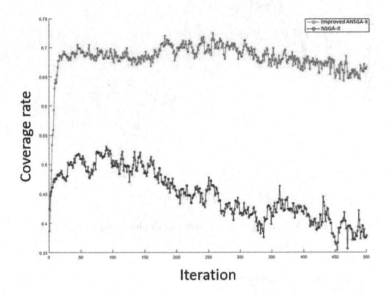

Fig. 4. Comparison of coverage rate.

4   Experimental Results

4.1   Experimental Setup

This experiment uses java 8 for encryption algorithm implementation and uses Intel i7 8700h processor as the main computing platform. In order to maintain high fairness, we use the original DES algorithm published on GitHub [2]. The existing work [1] used GA-based encryption which adopted the same hardware setting with this study as well as the GA parameter uniformly uses 12 size parent population and executes 16 rounds. Each round is executed 200 times. The selection operation of GA adopts tournament selection. The crossover and mutation triggering ratios are 95% and 50% respectively.

4.2   Results Analysis

In this study, MATLAB is adopted as a simulated tool to construct the proposed algorithms. In simulations, users and buildings are randomly deployed in our scenario of 1,000 m × 1,000 m. There are three types of small cells, called small cell 1, small cell 2, and small cell 3. They are divided according to cost and coverage.

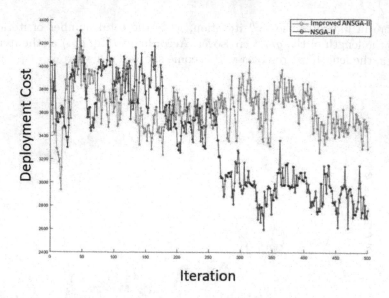

Fig. 5. Comparison of average deployment cost.

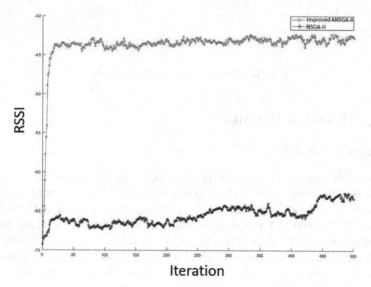

Fig. 6. Comparison of RSSI.

Macro cells and small cells form a wireless backhaul network in our scenario. To test the impact of the number of different users on the deployment performance, the number of users from 1,000 to 10,000 with 1,000 as a gap.

Figure 4, Fig. 5, and Fig. 6 are comparisons of coverage rate, deployment cost, and RSSI. The X-axis is the number of iterations and the Y-axis is three kinds of metrics. Falling into the local optimum. Compared with NSGA-II, although the deployment cost of this method is higher than that of NSGA-II, the RSSI and coverage are significantly better than NSGA-II. The reason is that when there are no candidate nodes, the solution is difficult to converge. For NSGA-II, although the deployment cost and RSSI have gradually improved, the coverage rate has also decreased significantly.

5   Conclusion

In this study, we formulated the 3D deployment problem as a MOO problem to optimize coverage, deployment cost, and RSSI. In addition, we proposed a three-dimensional small cell deployment algorithm based on NSGA-II. The simulation results show that although the deployment cost of the proposed method is higher than NSGA-II, it has better coverage and RSSI. It has a better effect on the balance of multi-objective problems.

Acknowledgment. This research was partly funded by the National Science Council of the R.O.C. under grants 108-2221-E-197 -012 -MY3 and MOST 107-2221-E-197-005-MY3 and 107-2221-E-259-005-MY3.

References

1. Lien, S.Y., Shieh, S.L., Huang, Y., Su, B., Hsu, Y.L., Wei, H.Y.: 5G new radio: waveform, frame structure, multiple access, and initial access. IEEE Commun. Mag. **55**(6), 64–71 (2017)
2. Kawasaki, K.: Millimeter wave transmission device, millimeter wave transmission method, and millimeter wave transmission system. U.S. Patent No. 9,608,683, 28 March 2017
3. Xiao, M., et al.: Millimeter wave communications for future mobile networks. IEEE J. Sel. Areas Commun. **35**(9), 1909–1935 (2017)
4. Akdeniz, M.R., et al.: Millimeter wave channel modeling and cellular capacity evaluation. IEEE J. Sel. Areas Commun. **32**(6), 1164–1179 (2014)
5. An, J., Yang, K., Wu, J., Ye, N., Guo, N., Liao, Z.: Achieving sustainable ultra-dense heterogeneous networks for 5G. IEEE Commun. Mag. **55**(12), 84–90 (2017)
6. Huang, C., Chen, Q., Tang, L.: Hybrid inter-cell interference management for ultra-dense heterogeneous network in 5G. Sci. China Inf. Sci. **59**(8), 082305 (2016)
7. Yang, G., Mao, X., Ge, M., Ding, Z., Yang, X.: On the energy-efficient deployment for ultra-dense heterogeneous networks with NLoS and LoS transmissions. IEEE Trans. Green Commun. Networking **2**(2), 369–384 (2018)
8. Ding, F., Zhang, D., Song, A., Li, J.: A coverage and repair optimization algorithm for hybrid sensor networks. J. Internet Technol. **19**(3), 909–917 (2018)
9. Nguyen, T.G., So-In, C., Nguyen, N.G.: Barrier coverage deployment algorithms for mobile sensor networks. J. Internet Technol. **18**(7), 1689–1699 (2017)
10. Wang, C., et al.: Cellular architecture and key technologies for 5G wireless communication networks. IEEE Commun. Mag. **52**(2), 122–130 (2014)
11. Amaldi, E., Capone, A., Malucelli, F.: Planning UMTS base station location: optimization models with power control and algorithms. IEEE Trans. Wirel. Commun. **2**(5), 939–952 (2003)

12. Zhao, W., Wang, S., Wang, C., Wu, X.: Approximation algorithms for cell planning in heterogeneous networks. IEEE Trans. Veh. Technol. **66**(2), 1561–1572 (2017)
13. Ghazzai, H., Yaacoub, E., Alouini, M., Dawy, Z., Abu-Dayya, A.: Optimized LTE cell planning with varying spatial and temporal user densities. IEEE Trans. Veh. Technol. **65**(3), 1575–1589 (2016)
14. Tseng, F.-H.: Network Planning for Type 1 and 1a Relay Nodes in LTE Network Planning for Type 1 and 1a Relay Nodes in LTE. Doctoral dissertation, Department of Computer Science and Information Engineering, National Central University. https://hdl.handle.net112962dmg77
15. 5G; Study on channel model for frequencies from 0.5 to 100 GHz (3GPP TR 38.901 version 14.3.0 Release 14). https:www.etsi.orgdeliveretsi138900_13899913890114. 03.00 60tr_138901v140300p.pdf
16. Deb, K., Pratap, A., Agarwal, S., Meyarivan, T.: A fast and elitist multiobjective genetic algorithm: NSGA-II. IEEE Trans. Evol. Comput. **6**(2), 182–197 (2002)

An Efficient Cost Performance Placement of Macro Sites and Small Cells Under Restricted Topology

Fan-Hsun Tseng[1]([✉]) [iD], Li-Der Chou[2], Chi-Yuan Chen[3] [iD], and Han-Chieh Chao[4] [iD]

[1] Department of Computer Science and Information Engineering, National Cheng Kung University, Tainan, Taiwan
[2] Department of Computer Science and Information Engineering, National Central University, Taoyuan, Taiwan
cld@csie.ncu.edu.tw
[3] Department of Computer Science and Information Engineering, National Ilan University, Yilan, Taiwan
chiyuan.chen@ieee.org
[4] Department of Electrical Engineering, National Dong Hwa University, Hualien, Taiwan
hcc@mail.ndhu.edu.tw

Abstract. The global COVID-19 pandemic leads people to intermittent quarantines and lockdowns. Many large and crowded gatherings were postponed or even cancelled to prevent social distance violation. The paper aims to tackle the placement problem of macro sites, microcells and picocells under a restricted network topology. The cell placement problem is defined based on linear programming. The algorithm named Cost Efficiency algorithm is proposed to construct a network with higher performance and lower cost. Simulation results showed that the proposed algorithm yields higher SINR value and more number of served users over construction cost compared with other planning algorithms. The result of this work is expected to help users have better network service quality when they are isolated in hospital or self-health monitoring at home.

Keywords: Cost performance index · Network planning · Relay network · Restricted network topology · Signal-to-interference-plus-noise ratio · Small cell

1 Introduction

Relay technique not only extends the signal coverage of macro site but also improves the communication quality of users in small cells. Relay network has been investigated in various mobile network technologies, such as mobile multi-hop relay network in IEEE 802.16j [1], relay nodes in LTE-Advanced network [2], and small cells in heterogeneous network (HetNet) [3]. Relay placement is complicated in mobile networks since small cells are heterogeneous to macro site [4].

In IEEE 802.16j, transparent and non-transparent relay stations are deployed to extend base station's coverage [5]. In LTE-Advanced relay network, different types of

© ICST Institute for Computer Sciences, Social Informatics and Telecommunications Engineering 2021
Published by Springer Nature Switzerland AG 2021. All Rights Reserved
T. Nakano (Ed.): BICT 2021, LNICST 403, pp. 15–25, 2021.
https://doi.org/10.1007/978-3-030-92163-7_2

relay nodes are utilized to strengthen and improve eNB's signal and communication range [6]. Type 1 inband relay nodes can be deployed to extend cell range [7]. Small cell can be classified into microcell, picocell and femtocell on the basis of cell's transmission range. In general, a micro site provides communication range around several kilometers and microcell's transmission radius is around half one kilometer. Without loss of generality, picocell and femtocell are less than hundred meters. However, network providers trade cell's communication range off against the construction cost [8].

In this paper, the placement problem of multiple micro sites, microcells and picocells under a restricted network topology is considered and solved. The planning problem of a HetNet is formulated based on linear programming. A novel planning algorithm named Cost Efficiency algorithm is proposed to maximize network performance such as the number of served users and the signal-to-interference-plus-noise ratio (SINR) value. A realistic planning case and simulation-based results are given to prove the technical contribution of this work. The main contributions of the paper are summarized in the following.

- The placement problem of multiple macro sites, microcells and picocells is formulated on the basis of graph theory and linear programming.
- The Cost Efficiency algorithm is proposed to maximize the ratio of SINR value and number of served users over construction cost.
- The planning result of a large-scale network as well as 900 km^2 contains several restricted areas is showed and discussed.
- Simulation-based results prove that the proposed Cost Efficiency algorithm outperforms other planning algorithms in terms of SINR value, the number of served users, network capacity, and construction cost.

The rest of the paper is organized as follows. Section 2 surveys the related works of relay technique in mobile networks. Section 3 introduces the network model and problem formulation. In Sect. 4, the proposed algorithm is explained. Simulation results are shown in Sect. 5. Section 6 concludes the paper and provides the future direction of this work.

2   Related Works

Small cell deployment has been widely studied in recent years [9], such as handover algorithm [10], coverage extension [11], safety relays [12]. In [13], the researchers proposed a soft frequency reuse scheme to enhance edge user coverage and improve network performance. The researchers in [14] utilized small cells to improve the signal dead zone. In [15], the researchers proposed backhaul traffic models to optimize the energy efficiency of small cell backhaul networks. However, it is a trade-off problem between network performance and construction cost [16]. It is difficult to achieve cost economized and performance improved at the same time [17].

The most relevant works to this paper are [18–20]. In [18], the Supergraph Tree algorithm was proposed to achieve construction cost economized. However, the Supergraph Tree algorithm cannot guarantee the network performance of planning result.

Therefore, the Set Covering algorithm in [19] was proposed to enhance the network performance. Although the Set Covering algorithm improves network capacity, it also raises construction cost. In [20], the Tree with Type 1 and Type 1a relay algorithm was proposed to eliminate communication interference and decrease construction cost. However, the Tree with Relay algorithm cannot achieve the highest ratio of network performance over cost. As a result, the Cost Efficiency algorithm is proposed to accomplish the trade-off problem between network performance and construction cost at the same time.

3   Problem Definition

The network model is formulated as an undirected graph based on tree structure. Let V be the set of vertex include candidate positions (CPs) and user equipment (UE), and E be the set of edges between vertex. Table 1 lists the definition of notations.

Table 1. Definition of notations

Variable	Definition
G	Planned field of interest
V_1	Set of CPs in G
m	Number of CPs in G, $m = |V_1|$
V_2	Set of UEs in G
n	Number of UEs in G, $n = |V_2|$
E_1	Set of links within V_1
E_2	Set of links between V_1 and V_2
c_i	Construction cost of deployed cell on CP
$x_{i,j}$	Available links within V_1
$a_{i,k}$	Available links within V_1 and V_2
$w_{i,k}$	Signal quality of a UE in a CP
$y_{i,k}$	Hop count limitation
E_1^l	Depth of the routing tree within l
δ	Minimum user utility required
η_i	Maximum depth l of routing tree

Given an undirected graph $G = (V, E)$, where $V = V_1 \cup V_2$ and $E = E_1 \cup E_2$. Let V_1 be the subset of CPs and V_2 be the subset of UEs. Let E_1 be the subset of links between CPs in V_1, and E_2 be the subset of link between CP and UE. If a cell is deployed on the CP corresponding to z_i, it is defined as

$$z_i = \begin{cases} z_{macro}, & \text{if macro site deployed} \\ z_{micro}, & \text{if microcell deployed} \\ z_{pico}, & \text{if picocell deployed} \\ 0, & \text{otherwise} \end{cases}, \forall i \in V_1. \tag{1}$$

Note that $z_i = 0$ when not deployment on CP z_i. Thus, $z_i \in \{0, z_{macro}, z_{micro}, z_{pico}\}$. The construction cost of different kinds of cells are different and defined as

$$c_i = \begin{cases} c_{macro}, \text{ if macro site deployed} \\ c_{micro}, \text{ if microcell deployed} \\ c_{pico}, \text{ if picocell deployed} \\ 0, \text{ otherwise} \end{cases}, \forall i \in V_1. \tag{2}$$

Note that $c_i = 0$ while not deploying cells on CP z_i. Thus, the construction cost $c_i \in \{0, c_{macro}, c_{micro}, c_{pico}\}$.

Let $x_{i,j}$ be the available link between CP z_i and CP z_j, where

$$x_{i,j} = \begin{cases} 1, \text{ if} (z_i, z_j) \in E_1 \text{and} z_i \cdot z_j \neq 0 \\ 0, \text{ otherwise} \end{cases}, \forall i, j \in V_1. \tag{3}$$

If $x_{i,j} = 1$, it means that the link between CP z_i and z_j is valid, and vice versa. To describe the coverage of a specified UE q_k, the variable $a_{i,k}$ is defined as

$$a_{i,k} = \begin{cases} 1, \text{ if } (z_i, q_k) \in E_2 \\ 0, \text{ if } (z_i, q_k) \notin E_2 \end{cases}, \forall i \in V_1, \forall k \in V_2. \tag{4}$$

If $a_{i,k} = 1$, it means that the link between CP z_i and UE q_k is valid, and vice versa. For receiving signal, a UE must be served by a specific cell at least. As a result, we assume that the UE q_k is directly served by a macro site or receives signal from microcell or picocell which relays from a micro site. The signal quality $w_{i,k}$ of UE q_k is defined as

$$w_{i,k} = \begin{cases} w_{macro}, \text{ if } a_{i,k} = 1 \wedge z_i = z_{macro} \\ w_{micro}, \text{ if } a_{i,k} \cdot x_{i,j} = 1 \wedge z_i = z_{micro} \\ w_{pico}, \text{ if } a_{i,k} \cdot x_{i,j} = 1 \wedge z_i = z_{pico} \\ 0, \text{ otherwise} \end{cases}, \tag{5}$$

where $\forall i, j \in V_1, \forall k \in V_2$. Therefore, the utility of a specified UE q_k is $u_k = max_k w_{i,k}$.

In addition, the tree structure should be restricted and formulated as

$$\sum_{e \in E(S)} x_e \leq |S| - 1. \tag{6}$$

where $S \subseteq V_1$ and $E(S) = \{(i, j) \in E_1 | i, j \in S\}$. Since two-hop relaying is considered in this work, the depth of routing tree should be defined as

$$y_{i,k} = \begin{cases} 1, \text{ if } (z_i, q_k) \in E_1^l \\ 0, \text{ if } (z_i, q_k) \notin E_1^l \end{cases}, \forall i \in V_1, \forall k \in V_2. \tag{7}$$

Let $\eta_i = max_j y_{i,k}$ and define η_i should be greater than or equal to 1 and less than or equal to 2.

The planning problem with considering multiple macro sites, microcells and picocells deployment for a HetNet is formulated as follows.

Maximize

$$\sum_{i=1}^{m} \sum_{k=1}^{n} \frac{z_i \cdot w_{i,k}}{c_i}, \tag{8}$$

Subject to

$$\sum_{i=1}^{m} z_i \geq 0, 0 \leq z_i \leq m, \forall i \in V_1, \tag{9}$$

$$\sum_{i=1}^{m} \sum_{k=1}^{n} w_{i,k} = a_{i,k} \cdot y_{i,k} \cdot w_{i,k}, \tag{10}$$

$$\sum_{i=1}^{m} c_i \leq m \times c_{macro}, \forall i \in V_1, \tag{11}$$

$$\sum_{e \in E(S)} x_e \leq |S| - 1. \tag{12}$$

$$\sum_{k=1}^{n} \frac{u_k}{n} \geq \delta, \forall k \in V_2. \tag{13}$$

$$1 \leq \eta_i \leq 2, \forall i \in E. \tag{14}$$

The primary goal is to maximize the ratio of served user and SINR value over construction cost, which is captured in Eq. (8). Constraint (9) guarantees micro sites and small cells are deployed. Constraint (10) guarantees users are served by cell deployed. Constraint (11) guarantees the construction cost of deployed cells not exceed the maximum construction cost while deploying macro sites on all CPs. Constraint (12) restricts the network topology to tree structure. Constraint (13) guarantees that all served users are provided with minimum utility rate required. Constraint (14) guarantees the planning result is two-hop relaying.

4 Proposed Cost Efficiency Algorithm

In this section, the Cost Efficiency algorithm is proposed to tackle with the placement of macro sites, microcells and picocells. The notations in the algorithm are explained as follows. A descending order list $S = f_d(G, X, Y)$ is used to generate the adjacent vertices in Y of elements in X. If vertexes are in same degree, the sequence can be arbitrary. The notation $S[i]$ represents the i-th vertex in the list S. The notation $N(G, x)$ represents the set of neighbor vertices of vertex x in graph G. The notation $\gamma(G, X)$ represents the vertex-induced subgraph of G with the complementary set $V \backslash X$. Let $BFS(G, x)$ be a subgraph roots at the vertex x and applies breath-first-search approach. The notation $U(z)$ represents the average utility of served users. The notation $f_{BFS, l, macro}(G)$ represents the graph constructed by adopting l-depth breadth-first-search on graph G from a deployed macro site.

The proposed Cost Efficiency algorithm is captured in Algorithm 1. Line 1 initializes the placement configuration. Line 2 to line 7 arranges CPs in a descending order list in accordance with the number of served UEs. Note that a selected CP and its served UEs forms a subgraph. Line 8 to line 13 deploys microcell on an unselected CP to connect the subgraphs. A CP connects the most other CPs will be selected to deploy a macro site when more than one CP are connected with each other. Line 14 to line 20 guarantees all served UEs are satisfied with their minimum utility rate. If the utility rate of a served UE is less than its minimum utility rate, a picocell will be deployed to replace the current deployed cell. Line 21 to line 26 guarantees the placement result follows tree structure and does not violate hop count limitation. Lastly, the placement result of a given network topology includes micro sites, microcells and picocells is obtained.

Algorithm 1: Cost Efficiency Algorithm

Input: G: *given undirected graph*
$\quad\quad\quad z$: *placement of macro site, microcell and picocell*
$\quad\quad\quad \delta$: *minimum user utility required*
$\quad\quad\quad l$: *maximum routing tree length*

01 $\Omega_1 = V_1, \Omega_2 = V_2, \Delta = \emptyset, l = 2$
02 **repeat**
03 $\quad S = f_d(G, \Omega_1, \Omega_2)$
04 $\quad \Delta = \Delta \cup S[1]$
05 $\quad \Omega_1 = \Omega_1 \backslash S[1] \ and \ \Omega_2 = \Omega_2 \backslash N(G, S[1])$
06 **until** $\Omega_2 = \emptyset$
07 $G_r = \gamma(G, \Omega_1), where \ G_r = (V_{G_r}, E_{G_r})$
08 **while** G_r *is unconnected*
09 $\quad S = f_d(G, \Omega_1, V_{G_r})$
10 $\quad z_{S[1]} = z_{micro}$
11 $\quad G_r = \gamma(G, \Omega_1 \backslash S[1])$
12 $\quad z_{V_{G_r}} = z_{macro}$
13 $\Omega_3 = \emptyset, G_{r'} = G_r$
14 **while** $V_{G_r} \neq \emptyset$ and $U(z) < \delta$
15 **repeat**
16 $\quad S = f_d(G_r, V_{G_r}, \Omega_2)$
17 $\quad z_{S[1]} = z_{pico}$
18 $\quad \Omega_3 = \Omega_3 \cup \{S[1]\}$
19 $\quad G_r = \gamma(G_r, S[1])$
20 **until** $U(z) \geq \delta$
21 **repeat**
22 $\quad G^l = f_{BFS,l,macro}(G_r), where \ G^l = (V^l, E^l)$
23 $\quad$ **if** $V^l \neq \Delta$
24 $\quad\quad S = f_d(G_r, \Delta \backslash V^l, V_2)$
25 $\quad\quad z_{S[1]} = z_{macro}$
26 **until** $V^l = \Delta$
27 **Output** the placement results of a given G

---

5   Simulation Results

Simulation is conducted by MATLAB [21] to construct network topology and planning algorithms. Parameters in simulation are listed in Table 2. Firstly, we assume that all locations of CPs and UEs are known so that algorithms decide to deploy cells on CPs for covering UEs. The range of micro site, microcell and picocell is set to 3, 0.5, 0.2 km respectively. According to [22, 23], the construction cost of micro site, microcell and picocell is defined as 20, 3 and 1 unit respectively. The simulation results are obtained and averaged from 10 different network topologies. The analyzed performance metrics include construction cost, number of served users, network capacity, and average SINR value of each user. The proposed Cost Efficiency algorithm is compared with three algorithms, i.e., the Supergraph Tree in [18], the Set Covering algorithm in [19], and the Tree with Relay algorithm in [20].

Table 2. Simulation parameters

Variables/Parameters	Value
Network size	30 × 30 (km)
Number of restricted zone	10
Number of candidate position	100
Number of user equipment	1000
Macro site's range radius	3 (km)
Microcell's range radius	500 (meter)
Picocell's range radius	200 (meter)
Macro site's construction cost	20 (unit)
Microcell's construction cost	3 (unit)
Picocell's construction cost	1 (unit)
Macro site's transmission power	46 (dBm)
Microcell's transmission power	32 (dBm)
Picocell's transmission power	15 (dBm)
Propagation model	−174 (dBm/Hz)
Noise power	2×10^{-14}(W)
Bandwidth	10 (MHz)

The average construction cost of four algorithms in 10 planning results is captured in Fig. 1. It can be observed that both of the Set Covering algorithm and the Tree with Relay algorithm spend higher construction cost compared with other algorithms. The Set Covering algorithm yields the highest construction cost because it aims to serve more users that results in the most macro site deployment. The Supergraph Tree algorithm aims to minimize construction cost but results in the lowest number of served users. Although the proposed Cost Efficiency algorithm spends more cost than the Supergraph Tree algorithm, it achieves significantly higher number of served users.

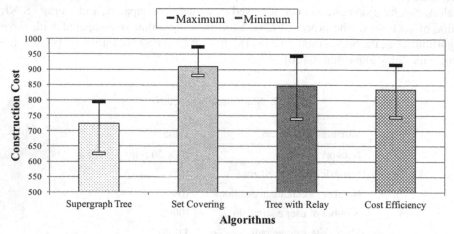

Fig. 1. Average construction cost of four algorithms.

The average number of served users in 10 planning results is captured in Fig. 2. It can be observed that the Set Covering algorithm and the proposed Cost Efficiency algorithm yield more served users compared with the other two algorithms. Although the Supergraph Tree algorithm economizes construction cost, it cannot guarantee more served users. Compared to the Set Covering algorithm, the Cost Efficiency algorithm achieves slightly fewer served users but lower construction cost significantly.

The average network capacity of four algorithms in 10 planning results is captured in Fig. 3. It can be observed that the Set Covering algorithm and the proposed Cost Efficiency algorithm achieve higher network capacity compared with the other two algorithms. The Supergraph Tree algorithm obtains lower network capacity because it serves the lowest number of served users. The Tree with Relay algorithm obtains the lower network capacity because it constructs the planning result with worse SINR value. The Cost Efficiency algorithm yields the similar network capacity with slightly less served users compared to the Set Covering algorithm. It is attributed to the fact that the Cost Efficiency algorithm serves users by deploying microcells and picocells with better SINR value for achieving higher network capacity.

The average SINR value of four algorithms in 10 planning results is captured in Fig. 4. It can be observed that the proposed Cost Efficiency algorithm achieves the highest SINR value compared with other algorithms. The Cost Efficiency algorithm deploys cheaper microcells and picocells to provide better communication quality so

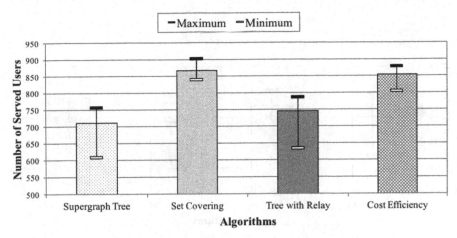

Fig. 2. Average number of served users in 10 planning results.

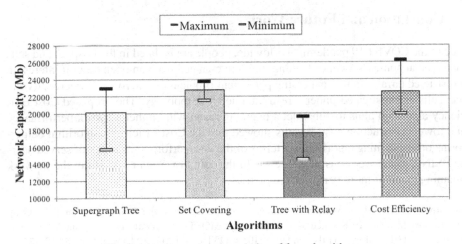

Fig. 3. Average network capacity of four algorithms.

that it accomplishes higher network capacity by serving more users with the highest SINR value. In summary, it not only uses lower construction cost to serve more users but also yields the best communication quality.

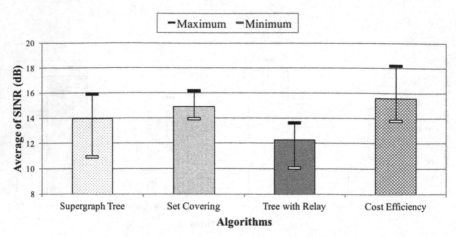

Fig. 4. Average SINR value of all served users.

6   Conclusion and Future Work

Due to the COVID-19 pandemic worldwide, people are isolated in the hospital and self-managed at home. Network planning problem changed from an open field of interest to a restricted terrain. In this paper, the placement problem of macro sites, microcells and picocells is investigated under a restricted network topology. The proposed Cost Efficiency algorithm aims to construct a planning result with higher network performance and lower construction cost. Results showed that the Cost Efficiency algorithm outperforms other planning algorithms in terms of the cost performance index of served users and SINR value over construction cost. In the future, we intend to study the network planning result of more performance metrics.

Acknowledgment. This work was financially supported from the Young Scholar Fellowship Program by Ministry of Science and Technology (MOST) in Taiwan, under Grant MOST109-2636-E-003-001, and was partly funded by the MOST in Taiwan, under grant MOST108-2221-E-008-033-MY3, MOST107-2221-E-197-005-MY3, and MOST 107-2221-E-259-005-MY3.

References

1. Chen, C.-Y., Tseng, F.-H., Lai, C.-F., Chao, H.-C.: Network planning for mobile multi-hop relay networks. Wirel. Commun. Mob. Comput. **15**(7), 1142–1154 (2015)
2. Tayyab, M., Koudouridis, G.P., Gelabert, X., Jäntti, R.: Uplink reference signal based handover with mobile relay node assisted user clustering. In: IEEE Global Communications Conference, pp. 1–6. IEEE, Taipei, Taiwan (2020)
3. Lahad, B., Ibrahim, M., Lahoud, S., Khawam, K., Martin, S.: Uplink/Downlink decou-pled access with dynamic TDD in 5G HetNets. In: 16th International Wireless Communications and Mobile Computing Conference, pp. 1330–1335. IEEE, Limassol, Cyprus (2020)
4. Tseng, F.-H., Chen, C.-Y., Chao, H.-C.: Multi-objective optimization for heterogeneous cellular network planning. IET Commun. **13**(3), 322–330 (2019)

5. Lara-Cueva, R., Custodio-Rivera, D., Benítez, D.S.: Performance analysis of uplink capacity in IEEE 802.16 transparent and non-transparent modes. In: 2th IEEE Second Ecuador Technical Chapters Meeting, pp. 1–5. IEEE, Salinas, Ecuador (2017)
6. Mahdi, Z.H., Yahiya, T.A., Kirci, P.: Scheduling algorithms comparison in HetNet based LTE-A. In: 3th International Symposium on Multidisciplinary Studies and Innovative Technologies, pp. 1–5. IEEE, Ankara, Turkey (2019)
7. Saleh, A.B., Bulakci, Ö., Redana, S., Hämäläinen, J.: On cell range extension in LTE-advanced type 1 inband relay networks. Wirel. Commun. Mob. Comput. 15(4), 770–786 (2015)
8. Machuca, C. Mas, Rahman, M., Grobe, K., Kellerer, W.: Cost savings dependence on base station inter-distance in converged access network planning of dense populated areas. In: 16th International Conference on Transparent Optical Networks, pp. 1–4. IEEE, Graz, Austria (2014)
9. Tayyab, M., Gelabert, X., Jäntti, R.: A simulation study on handover in LTE ultra-small cell deployment: A 5G challenge. In: 2th IEEE 5G World Forum, pp. 1–5. IEEE, Dresden, Germany (2019)
10. Huang, Z.-H., Hsu, Y.-L., Chang, P.-K., Tsai, M.-J.: Efficient handover algorithm in 5G networks using deep learning. In: IEEE Global Communications Conference, pp. 1–6. IEEE, Taipei, Taiwan (2020)
11. Cao, L., Yue, Y., Cai, Y., Zhang, Y.: A novel coverage optimization strategy for heterogeneous wireless sensor networks based on connectivity and reliability. IEEE Access 9, 18424–18442 (2021)
12. Sun, Y., Cao, Y., Zhang, Y., Xu, C.: A novel life prediction method for railway safety relays using degradation parameters. IEEE Intell. Transp. Syst. Mag. 10(3), 48–56 (2018)
13. Abbas, Z.H., Haroon, M.S., Muhammad, F., Abbas, G., Li, F.Y.: Enabling soft frequency reuse and stienen's cell partition in two-tier heterogeneous networks: cell deployment and coverage analysis. IEEE Trans. Veh. Technol. 70(1), 613–626 (2020)
14. Yamamoto, T., Konishi, S.: Impact of small cell deployments on mobility performance in LTE-Advanced systems. In: 24th IEEE International Symposium on Personal, Indoor and Mobile Radio Communications, pp. 189–193, IEEE, London, UK (2013)
15. Ge, X., Tu, S., Han, T., Li, Q., Mao, G.: Energy efficiency of small cell backhaul networks based on Gauss–Markov mobile models. IET Netw. 4(2), 158–167 (2015)
16. Nasri, R., Latrach, A., Affes, S.: Throughput and cost-efficient interference cancelation strategies for the downlink of spectrum-sharing Long term evolution heterogeneous networks. Wirel. Commun. Mob. Comput. 16(2), 236–248 (2016)
17. Teixeira, E.B., Ramos, A.R., Lourenço, M.S., Velez, F.J., Peha, J.M.: Capacity/cost trade-off for 5G small cell networks in the UHF and SHF bands. In: 22th International Symposium on Wireless Personal Multimedia Communications, pp. 1–6, Lisbon, Portugal (2019)
18. Tseng, F.-H., Chen, C.-Y., Chou, L.-D., Wu, T.-Y., Chao, H.-C.: A study on coverage problem of network planning in LTE-Advanced relay networks. In: 26th IEEE International Conference on Advanced Information Networking and Applications, pp. 944–950. IEEE, Fukuoka, Japan (2012)
19. Tseng, F.-H., Chou, L.-D., Chao, H.-C., Yu, W.-J.: Set cover problem of coverage planning in LTE-Advanced relay networks. Int. J. Electron. Commer. Stud. 5(2), 181–198 (2014)
20. Tseng, F.-H., Chou, L.-D., Chao, H.-C.: Network planning for type 1 and type 1a relay nodes in LTE-advanced networks. Wirel. Commun. Mob. Comput. 16(12), 1526–1536 (2016)
21. MATLAB: http://www.mathworks.com/products/matlab/ Accessed 12 Dec 2020
22. Lang, E., Redana, S., Raaf, B.: Business impact of relay deployment for coverage extension in 3GPP LTE-Advanced. In: IEEE International Conference on Communications, pp. 1–5. IEEE, Dresden, Germany (2009)
23. Khirallah, C., Thompson, J.S., Rashvand, H.: Energy and cost impacts of relay and femtocell deployments in long-term-evolution advanced. IET Commun. 5(18), 2617–2628 (2011)

Establishing an Integrated Push Notification System with Information Security Mechanism

Hsin-Te Wu[✉]

Department of Computer Science and Information Engineering, National Ilan University,
Yilan, Taiwan
hsinte@niu.edu.tw

Abstract. Today, many enterprises and governmental units utilize push technology to deliver messages. To enterprises, the system can announce company policies rapidly; to clients, it can distribute new events or promotions. There are many approaches to convey messages, such as emails and text and LINE messages; however, it requires a method to protect client information and avoid hackers or data theft by internal staff. The paper aims to develop a push notification system similar to a set-top box, and the system includes below features. 1. Establishing a set-top push system and server for firms to set login information. 2. Creating a function module for the operator to select suitable marketing models. 3. The integrated push notification system offers enterprises to send messages through emails, texts, and LINE application. 4. Through creating groups and user-friendly interfaces, operators could easily send push notices to various groups. 5. Build a hierarchical message authentication mechanism; the hierarchical method enables operators to review message content and ensure the correctness. 6. Develop personal information encryption and an Internet security mechanism to confirm the source, completeness, and authentication. 7. The personal information encryption protects the system from internal staff to export critical data. Clients will only need to rent the modules they need, and the supplier is in charge of providing the server and set-top push system, which will help to universalize the product.

Keywords: Natural language processing · Push technology · Network security · Personal information protection · Internet of Things

1 Introduction

With the development and application of mobile devices and mobile networks, information has become increasingly transparent. In the past, many marketers used flyers or letters for product promotion or event marketing information transmission. With the advent of virtual and real integration and the advent of the mobile era, many promotional activities are based on social Group software or SMS messaging allows customers to instantly grasp the latest promotions and make purchases through mobile phones. However, in the past, the cost of using SMS to push and broadcast was relatively high. For telecommunications manufacturers, a corporate newsletter cost 0.8 yuan. Small-scale e-commerce companies send large numbers of newsletters are a big burden, so

T. Nakano (Ed.): BICT 2021, LNICST 403, pp. 26–32, 2021.
https://doi.org/10.1007/978-3-030-92163-7_3

many companies nowadays use social software as the main message to push and broadcast. At present, many push broadcasts combine AI (Artificial Intelligence) for semantic analysis, and build dialog robots to understand customer needs based on client semantic analysis. The robots will answer customer questions according to needs, and collect customer-related questions for analysis, and understand the general customer's product. The frequently asked questions are improved, and the message push and broadcast combined with the dialogue robot effectively achieve the effect of 24-h customer service.

Nowadays, most people are accustomed to using social software for message transmission or information acquisition. Therefore, this paper's push broadcast system is added to Line's social software, and there is no upper limit for push messages. This paper mainly uses set-top boxes and module rentals. Therefore, price parity can help Taiwan's small and medium-sized enterprises or small e-commerce use. Message transmission is an important part of marketing. Therefore, companies can use this system to transmit the latest information to customers, and companies can classify customers and transmit marketing information according to customer groups. Due to the shortage of manpower for small and medium-sized enterprises, this system introduces artificial intelligence to provide customer service. In addition, this plan will classify customer problems one by one to help the company improve its products. At present, the security of customer data is very important, so this plan also introduces information security Mechanism to ensure the safety of customers' personal information, and internal personnel cannot obtain customer plaintext data from the server.

The methods proposed in this paper are mainly as follows: 1. Build an integrated push broadcast set-top box, and the equipment can be placed in the computer room for environmental monitoring. Each enterprise can set the set-top box account secret and organize the personnel account secret to ensure the data. Security, 2. In a functional modular way, companies can choose suitable functions. This plan uses the Internet of Things to build a set-top box to build a small exclusive business service station, and the data between companies will not circulate to ensure information. Security, 3. Build an integrated message push system. The message push methods include telephone, E-mail, SMS and Line, so that companies can choose the message push method, allowing the operator to push messages to SMS, In mail and Line, when some industries need urgent notification, they need to use the phone to notify relevant personnel to return to the company, or when the computer room environment is abnormal, they also need to notify. 4. Establish groups and simple UI interfaces to facilitate the group broadcast by operators. 5. Establish a hierarchical message content verification mechanism, through hierarchical personnel to view the message content, to ensure the correctness of the content, 6. Establish a network security mechanism to ensure the source, integrity and identity verification of the sent message, 7. Build a personal The data protection encryption mechanism prevents internal personnel from exporting customer personal data. The product customers of this plan only need to rent the required functions, and the server and set-top box are the responsibility of the manufacturer, which helps to popularize the product.

2 Related Works

In the literature [1], it is mainly mentioned that the Internet of Things is mainly based on real-time systems. Therefore, the overall system design must be lightweight to achieve real-timeness, and other systems need to add security mechanisms to ensure data security. . In the literature [2], it is mainly mentioned that the data of the Internet of Things is too large, so the Internet of Things needs to combine cloud computing or fog computing to store the data in a spatially penetrating storage before data processing or analysis can be performed. In the literature [3], VR-IOT is mainly proposed, in which IoT information transmission needs to rely on XML for mutual transmission, and MQTT is used for IoT device control. , The paper method information is sent to the server for backup. When the set-top box is damaged, the data can be imported from the server back to the new set-top box. In addition, when the set-top box module needs to be updated, the plan. If the XMPP system is installed, the module can be updated through commands, which can achieve the effect of system optimization.

In the literature [4], ID-based is mainly used for distributed data verification. As long as any party obtains the ID of the other party and then uses bilinear pairing to verify the legitimacy, the integrity and authentication of the data can be determined. He He 2015 mainly uses ID-based privacy and data security mechanisms for building ad hoc networks for vehicles. Through the ID-based mechanism, the anonymous ID of the vehicle can be effectively established, and when the vehicle commits a crime, the real ID of the vehicle can be passed. In the literature [5], bilinear pairing is mainly used to create data signatures, which can effectively verify the correctness of data. In the literature [6], bilinear pairing is mainly used to establish a security mechanism without public/private key authentication. Here, the time consumed by the public/private key authentication can be overcome and complete security can be provided. And personal asset protection.

3 The Proposed Scheme

3.1 Systme Model

The schematic diagram of the project system is shown in Fig. 1. The project system is mainly divided into a set-top box and a server side. The set-top box is mainly used as a small server by the development version of the Internet of Things, in which the set-top box can be connected to a sensor , Such as: temperature and humidity, flame sensors, etc., can also be connected to monitoring equipment to monitor the environment of the computer room, in addition, the set-top box can construct a small database for data storage, and the database has information security encryption to avoid exposure when data is stolen Customer's real data. The push broadcast system mainly has telephone, mail, SMS and line message communication functions. The set-top box has a simple network management protocol (SNMP) function. When the computer room environment is abnormal, the telephone will be activated to notify the network administrator. When there is an abnormality in the plant, plant personnel can notify the engineer to return to the plant through the push broadcast system. The line push broadcast system of this project mainly uses the Line ID left by the customer to join, and uses IFTTT

(If this, then that) to perform Message transmission. On the server side, there are mainly modules and software updates, databases, and XMPP platforms. Business owners can securely connect to the server side selection module through a set-top box. The server side will record which modules the enterprise uses. When new software needs to be updated, the set-top box can be notified of the update time through the XMPP platform. The owner can choose the time to update the software. The set-top box will actively download the latest module software from the server, and the data in the set-top box will be updated. Encrypted and backed up to the database regularly to avoid data loss due to damage to the set-top box.

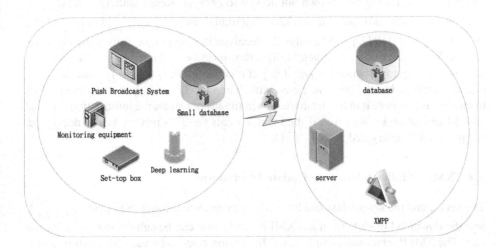

Fig. 1. XMPP system platform

3.2 Cybersecurity Mechanism Establishment

This article mainly uses Bilinear Pairings to construct an overall network security mechanism. Assuming TA is the server side, first calculate the safety factors such as the public key and Private key of the TA and the set-top box ($I_1 \sim I_n$). The calculation is as follows:

1. TA selects $c \in Z_q^*$ as the secret key, r is the public value.
2. The ID of TA is ID_{TA}, where public key is $\mathcal{PK}_{ID_{TA}} = ID_{TA} \cdot P$, and private key is $\mathcal{PR}_{ID_{TA}} = r^c \cdot ID_{TA} \cdot P$.
3. The public value of TA is $\mathcal{PU}_{ID_{TA}} = r^{\frac{1}{c}} \cdot P$.

Next, calculate the public key and private key of $I_1 \sim I_n$, which are calculated as follows:

1. The public key of I_n is $\mathcal{PK}_{ID_{I_n}} = ID_{I_n} \cdot P$.
2. The private key of I_n is $\mathcal{PR}_{ID_{I_n}} = r^c \cdot ID_{I_n} \cdot P$.

The public Key and private key of $I_1 \sim I_n$ above are the preset passwords of the set-top box. Among them ID_{I_1} is the name of the set-top box, and the company needs to change the private key when connecting for the first time.

3.3 Data Backup Mechanism

The machine-top box of this project transfers customer data and related data to the database. First, I_1 will use its own public key to encrypt. Reuse with TA $\mathcal{SK}'_{ID_{I_1 \leftrightarrow TA}}$ encrypt the data and send it to TA. Calculated as $\mathcal{SK}'_{ID_{I_1 \leftrightarrow TA}} (\mathcal{PK}_{ID_{I_1}} (M) \|$ $H(\mathcal{PK}_{ID_{I_1}} (M)))\| ID_{I_1}$. When the TA receives the ciphertext, it uses the common session key to decrypt it, and then verifies the integrity of the message. The TA uses hash technology to calculate $\mathcal{PK}_{ID_{I_1}} (M)$. If it is the $H\left(\mathcal{PK}_{ID_{I_1}} (M)\right)$ same, it means that the message has not been tampered with. Then the TA will use its own Public Key to encrypt it and store it in the database. When the insider steals the information, the real data M cannot be known, even if the insider steals the TA's privacy key is decrypted, but it cannot be decrypted $\mathcal{PK}_{ID_{I_1}} (M)$.

3.4 XMPP Platform Software Update Mechanism

The server and the set-top box can be securely connected through $SK\left(\mathcal{SK}'_{ID_{I_1 \leftrightarrow TA}}\right)$, and the download module from the XMPP server-side can be subscribed to the set-top box. The XMPP server can record the modules subscribed to by each set-top box, and is used to calculate the amount that the company needs to pay. When the XMPP server has a module to update the software, XMPP will issue a command to the IoT development board to notify the company of the update. The upper box will request the update time after receiving the instruction. At this time, the company must set the update time. If it is not set, it will update according to the preset time.

4 Experimental Result

The XMPP server of this project can generate reports, and can also control the related sensor equipment and related modules of the set-top box through commands, which will help the system to automate the process. 1. The system can confirm whether the enterprise subscription has paid for it, if there is a system then use the command to open the function, 2. The system can automatically determine which users need to update the software, and send out notifications by itself, as shown in Fig. 2 for the XMPP system platform.

Fig. 2. XMPP system platform

5 Conclusion

Many companies develop systems that are bought out or need to build hardware and software equipment. In addition, these systems need to be installed in servers. Therefore, the need for enterprises to set up servers is a burden for enterprises. The systems proposed in this project are mainly integrated push broadcasts. The system, combined with telephone, SMS, mail and Line, allows companies to choose the way of promotion. This plan also combines with an environmental monitoring system to help companies monitor the computer room or working environment. This plan uses the Internet of Things to increase the utilization rate of the enterprise. The set-top box version is built so that companies can use the set-top box to connect to the network and use it directly, reducing the incompatibility of the company for information operations. This plan adds an information security mechanism, and all customer data in the set-top box is encrypted. When internal or external personnel steal information, they cannot decrypt the data to achieve the security of customer data. The server built by this system manufacturer mainly provides data backup, module download and software update. Data backup is encrypted, so even if the server is If the device is stolen by insiders, the plaintext of the information cannot be obtained.

Acknowledgement. The authors would like to thank the anonymous reviewers for their valuable comments and suggestions on the paper. This work was supported in part by the Ministry of Science and Technology of Taiwan, R.O.C., under Contracts MOST 109-2622-E-197-012.

References

1. Condry, M.W., Nelson, C.B.: Using smart edge IoT devices for safer, rapid response with industry IoT control operations. Proc. IEEE **104**(5), 938–946 (2016)
2. Metzger, F., Hoßfeld, T., Bauer, A., Kounev, S., Heegaard, P.E.: Modeling of aggregated IoT traffic and its application to an IoT cloud. In: Proc. IEEE **107**(4), 679–694 (2019)
3. Simiscuka, A.A., Markande, T.M., Muntean, G.-M.: Real-virtual world device synchronization in a cloud-enabled social virtual reality iot network. IEEE Access **7**, 106588–106599 (2019)

4. Wang, H.: Identity based distributed provable data possession in multicloud storage. IEEE Trans. Serv. Comput. **8**(2), 328–340 (2015)
5. Tsai, J.-L.: A new efficient certificateless short signature scheme using bilinear pairings. IEEE Syst. J. **11**(4), 2395–2402 (2017)
6. Du, H., Du, H., Wen, Q.: A provably-secure outsourced revocable certificateless signature scheme without bilinear pairings. IEEE Access **6**, 73 846–73 855 (2018)

Double-Layered Cortical Learning Algorithm for Time-Series Prediction

Takeru Aoki$^{(\boxtimes)}$, Keiki Takadama, and Hiroyuki Sato

The University of Electro-Communications, 1-5-1 Chofugaoka, Chofu,
Tokyo 182-8585, Japan
{takeru-aoki,h.sato}@uec.ac.jp, keiki@inf.uec.ac.jp

Abstract. This work proposes a double-layered cortical learning algorithm. The cortical learning algorithm is a time-series prediction methodology inspired from the human neuro-cortex. The human neuro-cortex has a multi-layer structure, while the conventional cortical learning algorithm has a single layer structure. This work introduces a double-layered structure into the cortical learning algorithm. The first layer represents the input data and its context every time-step. The input data context presentation in the first layer is transferred to the second layer, and it is represented in the second layer as an abstract representation. Also, the abstract prediction in the second layer is reflected to the first layer to modify and enhance the prediction in the first layer. The experimental results show that the proposed double-layered cortical learning algorithm achieves higher prediction accuracy than the conventional single-layered cortical learning algorithms and the recurrent neural networks with the long short-term memory on several artificial time-series data.

Keywords: Cortical learning algorithm · Hierarchical temporal memory · Time-series data prediction

1 Introduction

Cortical learning algorithm (CLA) is a computational methodology based on hierarchical temporal memory [1–3], which is inspired by the human neuro-cortex. CLA is particularly suited to predict time-series data stream [4–6]. As related works on time-series prediction, the recurrent neural network (RNN) [7,8] with the long short-term memory (LSTM) [9] has been known as a representative and effective approach. It has been reported that CLA achieved higher prediction accuracy than LSTM on the taxi demand prediction task [10], and CLA is one of the promising time-series prediction algorithms.

The entire CLA predictor is called the *region*. The region is composed of multiple *columns*, and each column is composed of multiple *cells*. Each column has

T. Nakano (Ed.): BICT 2021, LNICST 403, pp. 33–44, 2021.
https://doi.org/10.1007/978-3-030-92163-7_4

column-synapses associating with input data bits. CLA uses column-synapses and encodes the input data value into internal data representation every time-step by making a subset of columns the *active* state. Also, CLA represents the input data context by making a subset of cells the *active* state and the input data incoming next time-step by making a subset of cells the *predictive* state. The conventional CLA predictor is composed of a single layer of one region. On the other hand, it has been known that the cerebral neocortex is composed of multiple layers. Since the multi-layering enhances the abstraction of internal data representation, we can expect the improvement of the prediction accuracy and the speeding up of the learning process in CLA.

In this work, we propose a double-layered CLA. Each layer is the conventional CLA predictor, and we propose interactions between layers. The first layer receives input data and transfers its data context representation to the second layer. The second layer makes a prediction based on the data context representation received from the first layer and makes feedback to the first layer. The first layer enhances the prediction based on the feedback from the second layer. We use multiple artificial time-series data for the experiments to verify the prediction performance of the proposed double-layered CLA. We have conducted a preliminary attempt on multi-layer CLA that tried some prototypes on an artificial time-series prediction [11]. The proposed double-layered CLA in this paper does not emphasize synapse connections for the first layers' active cells predicted by the second layer for stability. Also, the proposed double-layered CLA in this paper relaxes the condition to make cells the predictive state in the first layer with a parameter α, and the appropriate α is found out experimentally. The effectiveness of the proposed method is verified on several artificial time-series prediction tasks while comparing with simple conventional CLA and ANN-based LSTMs.

2   Cortical Learning Algorithm

2.1   Predictor Structure

Figure 1 shows the conventional structure of CLA predictor [4–6]. CLA receives an input data bit-string $x = (x_1, x_2, \ldots, x_n)$ every time-step t. The entire CLA predictor is called the region, which is composed of n_c columns c_i ($i = 1, 2, \ldots, n_c$). Figure 1 shows the case with $n_c = 5$ columns.

Each column c_i has two states: *normal* and *active*. Each column has n_{cy} column-synapses $c_i.y_k$ ($k = 1, 2, \ldots, n_{cy}$), which construct the relation with input data bits. Each column-synapse $c_i.y_k$ has permanence value $c_i.y_k.p$, which determines the connection or the disconnection of the synapse. Column-synapse $c_i.y_k$ is connected when $c_i.y_k.p \geq \theta_c$ and disconnected when $c_i.y_k.p < \theta_c$, where θ_c is the connection threshold of column-synapses.

Each column c_i involves n_r cells $r_{i,j}$ ($j = 1, 2, \ldots, n_r$). Figure 1 shows the case with $n_r = 5$ cells. Each cell has three states: *normal*, *active*, and *predictive*. Also, each cell $r_{i,j}$ has cell-synapses $r_{i,j}.y_k$ ($k = 1, 2, \ldots$), which construct relations with other cells. Each cell-synapse $r_{i,j}.y_k$ has permanence value $r_{i,j}.y_k.p$.

Cell-synapse $r_{i,j}.y_k$ is connected when $r_{i,j}.y_k.p \geq \theta_r$ and disconnected when $r_{i,j}.y_k.p < \theta_r$, where θ_r is the connection threshold of cell-synapses.

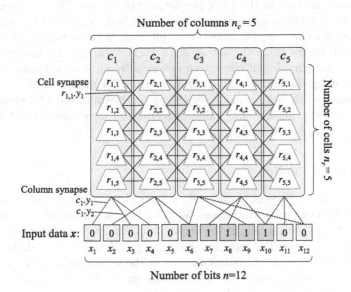

Fig. 1. CLA predictor

2.2   Algorithm

CLA repeats the following procedure every time-step t.

- Data encoding (Sect. 2.3)
- Spatial pooling (Sect. 2.4)
- Temporal pooling (Sect. 2.5)
- Data decoding (Sect. 2.6)

Details of each procedure are described as follows.

2.3   Data Encoding

CLA treats binary bit-string as input data value every time-step t. For a real value input data $X(t) \in [X^{\min}, X^{\max}]$ received at time-step t, CLA converts it to a binary bit-string $\boldsymbol{x} = (x_1, x_2, \ldots, x_n) \in \{0,1\}^n$ by using a chunk encoding method [6], which converts a real value into a bit-string involving continuous w chunk bits of 1. Figure 1 shows the case with the total bit length $n = 12$ and the chunk bit length $w = 5$.

2.4   Spatial Pooling

Spatial pooling is the process to convert the input data bit-string x into an internal representation in the CLA predictor. Specifically, CLA converts x into an active state combination of columns. CLA finds n_{ac} columns with many column-synapses connected to ones on the input data bit-string x and low active frequency and makes them the active state. CLA then updates the permanence value of column-synapses of the active columns to emphasize the internal data representation by the active pattern of columns. CLA increases the permanence value of column-synapses associated with one on the input data bit-string by p_c^+. CLA decreases the permanence value of column-synapses associated with zero on the input data bit-string by p_c^-.

2.5   Temporal Pooling

Temporal pooling represents input data context by an active state combination of cells and data incoming next time-step $t+1$ by a predictive state combination of cells. CLA first makes cells in the active columns the active state to represent the input data context. For each active column, CLA makes a predictive cell the active state if it exists. CLA makes all cells in the active column the active state otherwise. CLA then updates the permanence values of cell-synapses associated with the active cells transited from the predictive state to emphasize the cell-synapse network that succeeds the prediction. For each active cell, CLA increases the permanence value of cell-synapses associated with active cells at the previous time-step $t-1$ by p_r^+. Also, CLA decreases the permanence value of cell-synapses associated with non-active cells at the previous time-step $t-1$ by p_r^-.

Next, CLA makes cells with more than n_{ar} cell-synapses connected to the active cells at the current time-step t the predictive state. In this way, CLA internally represents the input data incoming next time-step $t+1$ as a combination of cells in the predictive state.

2.6   Data Decoding

CLA decodes the combination of cells in the predictive state to the data format same as the input data bit-string. In this work, we use the sparse distributed representations classifier (SDRC) [3] for the data decoding.

3   Proposal: Double-Layered Cortical Learning Algorithm

3.1   Predictor Structure

Figure 2 shows the predictor structure of the double-layered CLA we propose in this work. In the proposed double-layered CLA, the region of the conventional CLA shown in Fig. 1 is accumulated. The lower layer is called the first layer, and the higher layer is called the second layer. For the illustration, although cells in the first layer are positioned side-by-side in this figure, the predictor structures in the first and second layers are the same.

The first and second layers are connected by interlayer-synapses, which are column-synapses in the second layer. Each interlayer-synapse associates a cell in the first layer with a column in the second layer. The interlayer-synapses transfer cell information from the first layer to the second layer. Conversely, the interlayer-synapses transfer predictive information from the second layer to the first layer. In the double-layered CLA, the role of the first layer is similar to the conventional CLA, and the second layer affects to cell states in the first layer.

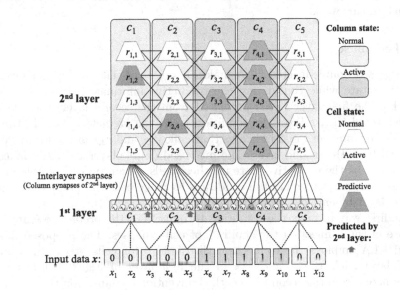

Fig. 2. Proposed double-layered CLA predictor

3.2   Algorithm

The first layer receives the input data x every time-step t. In the first layer, the spatial pooling makes a subset of columns the active state, the temporal pooling makes a subset of cells the active state. They are the same manner as the conventional CLA. Figure 2 show an example that columns c_3 and c_4 become the active state and represent the input data x in the first layer as the result of the spatial pooling. Also, cells $r_{3,2}$ and $r_{4,2}$ become the active state and represent the data context in the first layer as the result of the temporal pooling. Then, the first layer transfers the active cell pattern to the second layer through interlayer-synapses. In this process, the normal cell in the first layer is treated as 0, the active cell in the first layer is treated as 1, and the second layer performs the spatial pooling in the same manner as the conventional CLA. That is, the second layer makes a subset of columns the active state to represent the active cell pattern in the first layer. In the example of Fig. 2, we set 1 to cells

$r_{3,2}$ and $r_{4,2}$ and 0 to other cells in the first layer. The second layer receives them, and columns c_3 and c_4 become the active state in the second layer. Then, the second layer also performs the temporal pooling that makes a subset of cells in the second layer the active state and another subset of cells in the second layer the predictive state in the same manner as the conventional CLA. In the example of Fig. 2, cells of columns c_3 and c_4 in the second layer becomes the active state and makes cells $r_{1,2}$ and $r_{2,4}$ the predictive state in the second layer. Next, the second layer transfers the predictive cell information to the first layer through the interlayer-synapses, and the first layer determines predictive cells while considering the prediction from the second layer. This process is called the feedback in this work.

3.3   Feedback

The proposed double-layered CLA makes a subset of cells in the first layer the predictive state while considering the predictive information from the second layer. For the first layer's cells predicted from the second layer, the proposed double-layered CLA relaxes the condition to be the predictive state in the first layer. The example of Fig. 2 indicates that cells $r_{1,5}$ and $r_{2,5}$ in the first layer are predicted from the second layer. The proposed double-layered CLA relaxes the condition to be the predictive state for these cells. The proposed double-layered CLA uses two ways to relax the condition to make cells the predictive state in the first layer by employing a relaxing parameter $\alpha = [0,1]$.

The first way to relax the condition to make cells the predictive state is the decrease of the connection threshold θ_r of cell-synapses. The proposed double-layered CLA employs the connection threshold 0 for cells predicted from the second layer such as $r_{1,5}$ and $r_{2,5}$ in the first layer of Fig. 2. For other cells not predicted from the second layer, the conventional connection threshold θ_r is used. Since each cell must have more than n_{ac} cell-synapses connected to active cells to be the predictive state, the decrease of the connection threshold to 0 contributes to increasing the possibility to be the predictive state by increasing the number of connected cell-synapses.

The second way to relax the condition to make cells the predictive state is the decrease of the threshold to be the predictive state more directly. That is, the proposed double-layered CLA reduces the number of cell-synapses n_{ar} connected to active cells to be the predictive state. The proposed double-layered CLA employs the number of cell-synapses $\alpha \cdot n_{ar}$ connected to active cells for cells predicted from the second layer such as $r_{1,5}$ and $r_{2,5}$ in the first layer of Fig. 2. For other cells not predicted from the second layer, the conventional n_{ar} is used. This helps to make cells predicted from the second layer the predictive state in the first layer even if the cells do not have more than n_{ar} connected cell-synapses to active cells.

$\alpha = 1$ indicates that any prediction information from the second layer is not reflected in the first layer and brings the same results as the conventional single-layered CLA. Influence of the second layer increases by decreasing the relaxing parameter α from 1.

3.4 Interlayer-Synapse Arrangement

Each interlayer-synapse, column-synapse of the second layer, associates a column in the second layer with a cell in the first layer. The proposed double-layered CLA dynamically arranges interlayer-synapses over time. Concretely, during the spatial pooling in the second layer, we arrange interlayer-synapses associating second-layer columns with a low active frequency and active cells in the first layer when the number of active columns in the second layer is less than n_{ac}.

The dynamic interlayer-synapse arrangement starts after 10^4 time-steps since it cannot construct a valuable network until the lower layer is unstable.

3.5 Expected Effects

In the proposed double-layered CLA, we can expect to improve the learning speed of CLA since the feedback improves the learning efficiency. Also, we can expect to improve the prediction accuracy since the feedback from the second layer modifies the prediction in the first layer even if the prediction in the first layer does not work well.

4 Experimental Settings

4.1 Algorithms

To verify the effectiveness of the proposed CLA, we compared the conventional CLA [6], the simple CLA [12], the proposed CLA, and three RNN-based LSTMs using Adam, RMSprop, and SGD, which are network optimization algorithms.

4.2 Benchmark Time-Series Data

As benchmark time-series data, we use sine $X_s(t)$ and combined sine $X_c(t, m)$ as cyclic time-series data. Also, we use logistic mapping based $X_l(t)$ as non-cyclic time-series data. These benchmark time-series data are defined as follows:

$$X_s(t) = \frac{1}{2} \cdot \sin\left(\frac{(t-1) \cdot \pi}{50}\right) + \frac{1}{2}, \tag{1}$$

$$X_c(t, m) = \frac{1}{2} \cdot \sum_{k=1}^{m} \frac{1}{2k-1} \cdot \sin\left(\frac{(t-1) \cdot (2k-1) \cdot \pi}{50}\right) + \frac{1}{2}, \tag{2}$$

$$X_l(t) = \begin{cases} 0.4, & \text{if } t = 1, \\ 3.6 \cdot X_l(t-1) \cdot \{1 - X_l(t-1)\}, & \text{otherwise.} \end{cases} \tag{3}$$

For combined sine $X_c(t, m)$, we use $m = \{2, 3, 4, 5\}$. The range of time-steps is set to $t \in [1, 10^5]$. Input value range is set to $[X^{\min}, X^{\max}] = [-0.01, 1.01]$.

Table 1. Parameters of CLA in experiment

Name	Value
Length of the data bit-string n	421
Chunk length w	21
Number of columns n_c	2048
Number of column-synapses n_{cy}	22
Connection threshold of the column-synapses θ_c	0.1
Number of active columns n_{ac}	40
Increase permanence value for column synapses p_c^+	0.05
Decrease permanence value for column synapses p_c^-	0.025225
Number of cells in each column n_r in the first layer	4
Number of cells in each column n_r in the second layer	5
Connection threshold of the cell-synapses θ_r	0.5
Number of connected active cells to be the predictive state n_{ar}	15
Increase permanence value for cell synapses p_r^+	0.1
Decrease permanence value for cell synapses p_r^-	0.1

4.3   Metric

As a metric to assess the prediction accuracy of the time-series data, we use the prediction error $e(t)$ given by

$$e(t) = \sum_{\tau=t-99}^{t} |\bar{X}(\tau+1) - X(\tau+1)|, \qquad (4)$$

where, $\bar{X}(t+1)$ is the prediction value for the next time-step $t+1$, $X(t+1)$ is the true value at the next time-step $t+1$. Thus, $e(t)$ is the summation of the difference between the prediction and the true values every 100 time-steps. The smaller $e(t)$, the better prediction accuracy. In this work, we calculate $e(t)$ for $t \in \{100, 200, 300, 400, \ldots, 10^5\}$ in this work.

4.4   Parameters

As the implementation of the proposed CLAs, we employ the simple CLA, which involves the deterministic arrangement of the initial column-synapses, the fixed initial permanence value, and the aggregation of the column-synapse arrangement. For the three CLAs, the length of the data bit-string is set to $n = 421$ bits, and the chunk length is set to $w = 21$ bits. For the spatial pooling, the number of columns is set to $n_c = 2048$, the number of column-synapses is set to $n_{cy} = 22$. The connection threshold of the column-synapses is set to $\theta_c = 0.1$, the number of active columns at each time-step is set to $n_{ac} = 40$, the increase and the decrease permanence values are respectively $p_c^+ = 0.05$ and $p_c^- = 0.025225$.

For the temporal pooling, the number of cells in each column is set to $n_r = 4$ and 5 for the first and the second layers, respectively. The connection threshold of the cell-synapses is set to $\theta_r = 0.5$, the number of connected active cells to be the predictive state is set to $n_{ar} = 15$. Also, the increase and the decrease permanence values are respectively $p_r^+ = 0.1$ and $p_r^- = 0.1$. In the proposed double-layered CLA, we use the above parameters for both layers. Table 1 shows the parameters of CLA used in experiment.

We employ three LSTMs with different network optimization methods, Adam [13], RMSprop [14], and SGD. The Keras implementation [15] is employed in this paper. The neural network in the first layer is densely connected. The second layer is constructed as the LSTM with a tanh activation function. The output layer also has the densely connected neural network. For the loss function, the mean-squared error is used.

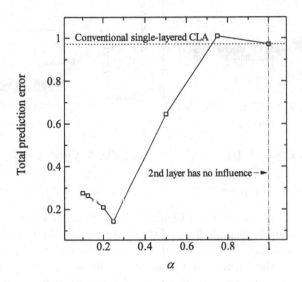

Fig. 3. Effects of parameter α to reflect the prediction from 2nd layer to the 1st layer on the combined sine $X_c(t, 2)$

5   Experimental Results and Discussion

5.1   Effects of Relaxing Parameter α

The proposed double-layered CLA has the relaxing parameter α to reflect the prediction in the second layer to the prediction in the first layer. The smaller α, the stronger influence from the second layer to the first layer.

Figure 3 shows the total prediction error of the proposed double-layered CLA on the combined sine $X_c(t, 2)$ in the last half of the entire time-steps when we vary the relaxing parameter α. $\alpha = 1$ indicates that the any prediction in the second layer does not reflect to the first layer. As a reference, the result of the conventional single-layered CLA is plotted as a horizontal dot line in this figure. From the result, we see that the total prediction error decreases by decreasing α from 1. We see that there is the best parameter α^* minimizing the total prediction error. In this case, $\alpha^* = 0.25$. Further decrease of α from α^* increases the prediction errors since the influence from the second layer becomes too strong for the first layer. In the following experiments, we use $\alpha^* = 0.25$ for the proposed double-layered CLA.

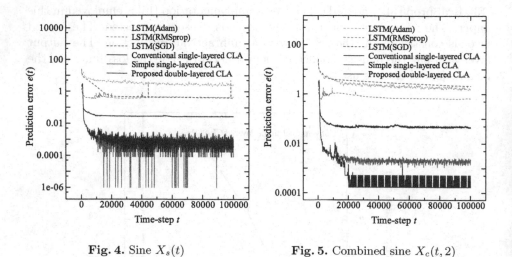

Fig. 4. Sine $X_s(t)$ **Fig. 5.** Combined sine $X_c(t,2)$

5.2 Prediction Accuracy

Figures 4, 5, 6, 7, 8, 9 show the histories of the prediction errors of the three LSTMs, the conventional single-layered CLA, the simple single-layered CLA and the proposed double-layered CLA on different benchmark time-series data.

Figure 4 shows the results on the most simple sin $X_s(t)$. In this case, we see that the prediction errors of the three CLAs are lower than those of the three LSTMs. In addition, we see that the simple CLA and the proposed CLA achieves lower prediction errors than the conventional CLA. On the other hand, we cannot see a clear difference between the simple single-layered CLA and the proposed double-layered CLA. However, we see that the cyclic increase of prediction error, especially in the case of the simple single-layered CLA. The proposed double-layered CLA suppresses the cyclic increase of the prediction errors on sin $X_s(t)$. This result reveals that the proposed double-layered CLA improves the prediction stability.

Figure 5 shows the results on the combined sine $X_c(t,2)$. From the result, we see that the proposed double-layered CLA significantly decreases the prediction errors compared with the conventional single-layered CLAs and LSTMs. Figures 6, 7, 8 respectively show results on the combined sine $X_c(t,3)$, $X_c(t,4)$, and $X_c(t,5)$. We see that the proposed double-layered CLA achieves lower prediction

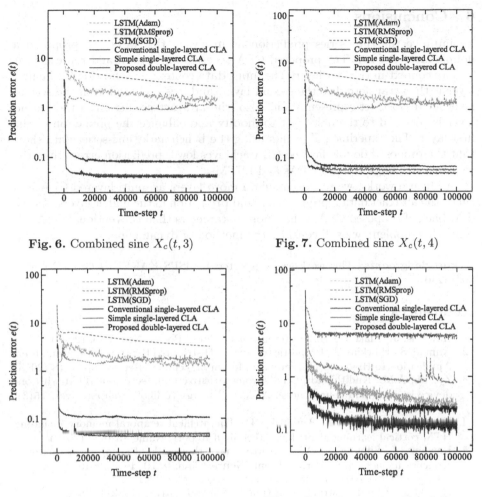

Fig. 6. Combined sine $X_c(t, 3)$

Fig. 7. Combined sine $X_c(t, 4)$

Fig. 8. Combined sine $X_c(t, 5)$

Fig. 9. Logistic mapping $X_l(t, 3.6)$

errors than the conventional single-layered CLAs and LSTMs. Figure 9 show the result on the time-series data on the logistic mapping. Also, in this case, we see a tendency that the proposed double-layered CLA achieves lower prediction errors than the conventional single-layered CLAs and LSTMs.

These results reveal that the double-layered CLA improves the prediction accuracy compared with the conventional single-layered CLA by accumulating the predictors of CLA and interacting between them based on the proposed interaction mechanism.

6 Conclusions

To improve the time-series prediction accuracy, in this work, we proposed a double-layered CLA. The proposed CLA receives input data and represents it and its context in the first layer. The input data context presentation in the first layer is then transferred to the second layer, and it is represented in the second layer as an abstract representation. Also, the abstract prediction in the second layer is reflected to the first layer to modify and enhance the prediction in the first layer. The experimental results using the benchmark time-series data show that the proposed double-layered CLA achieves lower prediction errors than the conventional single-layered CLAs and LSTMs.

As future works, we will accumulate more layers as multi-layered CLA and check the abstract data representation and prediction in the higher layers. As the drawback of proposed CLA, it has more parameters than conventional CLA. To solve this problem, we will consider the method of dynamic parameter setting.

Acknowledgments. This work was supported by JSPS KAKENHI Grant Number 20J14240.

References

1. Hawkins, J., Blakeslee, S.: On Intelligence. Times Books, New York (2004)
2. Ahmad, S., Hawkins, J.: Properties of Sparse Distributed Representations and their Application to Hierarchical Temporal Memory. Technical Report, pp. 1–18 (2015)
3. Zyarah, A.M., Kudithipudi, D.: Neuromemrisitive architecture of HTM with on-device learning and neurogenesis. ACM J. Emerg. Technol. Comput. Syst. **15**(3), 24 (2019). Article 24
4. Hawkins, J., Subutai, A., Dubinsky, D.: Hierarchical temporal memory including HTM cortical learning algorithms. Technical Report, Numenta Inc., (2010)
5. Hawkins, J., Subutai, A.: Why neurons have thousands of synapses, a theory of sequence memory in neocortex. Front. Neural Circuits **10**, 1–13 (2016)
6. https://github.com/numenta/nupic . Accessed 1 Mar 2021
7. Elman, J.L.: Finding structure in time. Cogn. Sci. **14**(2), 179–211 (1990)
8. Connor, J.T., Martin, R.D., Atlas, L.E.: Recurrent neural networks and robust time series prediction. IEEE Trans. Neural Networks **5**(2), 240–254 (1994)
9. Hochreiter, S., Schmidhuber, J.: Long short-term memory. Neural Comput. **9**(8), 1735–1780 (1997)
10. Cui, Y., Ahmad, S., Hawkins, J.: Continuous online sequence learning with an unsupervised neural network model. Neural Comput. **28**(11), 2474–2504 (2016)
11. Aoki, T., Takadama, K., Sato, H.: A preliminary study on a multi-layered cortical learning algorithm. In: The 7th UEC Seminar in ASEAN, 2020 and The 2nd ASEAN-UEC Workshop on Energy and AI (2020)
12. Aoki, T., Takadama, K., Sato, H.: Study on simple cortical learning algorithm and prediction accuracy improvement. In: SSI 2017, The Society of Instrument and Control Engineers, pp. 135–140 (2017) (in Japanese)
13. Kingma, D.P., Ba, J.: Adam: A Method for Stochastic Optimization. CoRR arXiv:1412.6980 (2014)
14. Hornik, K.: Approximation capabilities of multilayer feedforward networks. Neural Network **4**(2), 251–257 (1991)
15. https://github.com/keras-team/keras. Accessed 30 June 2021

Multi-factorial Evolutionary Algorithm Using Objective Similarity Based Parent Selection

Shio Kawakami$^{(\boxtimes)}$, Keiki Takadama, and Hiroyuki Sato

The University of Electro-Communications, 1-5-1 Chofugaoka,
Chofu, Tokyo 182-8585, Japan
{s.kawakami,h.sato}@uec.ac.jp, keiki@inf.uec.ac.jp

Abstract. This work proposes a multi-factorial evolutionary algorithm encouraging crossovers among solutions with similar target objective functions and suppressing crossovers among solutions with dissimilar target objective functions. Evolutionary multi-factorial optimization simultaneously optimizes multiple objective functions with a single population, a solution set. Each solution has a target objective function, and sharing solution resources in one population enhances the simultaneous search for multiple objective functions. However, the conventional multi-factorial evolutionary algorithm does not consider similarities among objective functions. As a result, solutions with dissimilar target objectives are crossed, and it deteriorates the search efficiency. The proposed algorithm estimates objective similarities based on search directions of solution subsets with different target objective functions in the design variable space. The proposed algorithm then encourages crossovers among solutions with similar target objectives and suppresses crossovers among solutions with dissimilar objectives. Experimental results using multi-factorial distance minimization problems show the proposed algorithm achieves higher search performance than the conventional evolutionary single-objective optimization and multi-factorial optimization.

Keywords: Multi-factorial optimization · Evolutionary algorithms · Objective function similarity

1 Introduction

In product developments and releases, multiple variant products with different specifications are often developed and released simultaneously. For instance, multiple smartphone models with different display sizes and processors depending on various users' demands are simultaneously developed and released from a device maker. These multiple models involve common parts that can be shared, and the independent and parallel design optimization of each model will not be efficient. This kind of design optimization problem belongs to the class of *multi-factorial optimization problem*. Multi-factorial optimization is that simultaneous

© ICST Institute for Computer Sciences, Social Informatics and Telecommunications Engineering 2021
Published by Springer Nature Switzerland AG 2021. All Rights Reserved
T. Nakano (Ed.): BICT 2021, LNICST 403, pp. 45–60, 2021.
https://doi.org/10.1007/978-3-030-92163-7_5

optimizations of multiple objective functions on common design variables. Evolutionary algorithms are particularly suited for multi-factorial optimization since solutions can share design variable information while parallelly optimizing multiple objective functions [1,2].

The conventional evolutionary approach [1–4] assigns an objective function to each solution as a *skill factor* and simultaneously optimizes multiple objective functions by using a single population, a solution set. To generate new solutions, the conventional MFEA [1,2] uses crossover, which recombines design variable values of solutions in the population. MFDE [3] is based on the differential evolution [5] and uses differential vectors of solutions in the design variable space. MFPSO [4] is based on the particle swarm optimizer [6] and uses velocity vectors obtained by historical and current solutions in the design variable space. These conventional algorithms commonly recombine solutions to generate new solutions. However, the conventional algorithms recombine solutions without considering similarities among objective functions. The recombinations of solutions with skill factors of similar objective functions encourage optimizations of their objective functions. However, the recombinations of solutions with skill factors of dissimilar objective functions deteriorate the simultaneous optimization of their objective functions. The use of similarities among objective functions for the recombinations would enhance evolutionary multi-factorial optimization.

To accelerate evolutionary multi-factorial optimization, in this work, we extend the conventional MFEA that uses the crossover for the recombination. The proposed algorithm estimates similarities among objective functions and utilizes them for crossovers to generate new solutions. The proposed algorithm encourages crossovers of solutions with skill factors of similar objective functions. This enhances the information exchange among these solutions and synergic effects for optimizations of these similar objective functions. Conversely, the proposed algorithm suppresses crossovers of solutions with skill factors of dissimilar objective functions. We use multi-factorial distance minimization problems [7] with several situations of objective function similarities and compare the search performance of the proposed algorithm with the conventional MFEA [1,2] and the single-objective EA without any interactions of solutions with different skill factors.

2   Multi-factorial Optimization Problem

2.1   General Definition

The general definition of multi-factorial optimization problems is as follows. For design variable vector $x = (x_1, x_2, \ldots, x_d)$ in the design variable space $\mathcal{X}$, there are K $(= |\mathcal{K}|)$ objective functions f_k $(k \in \mathcal{K})$, where $\mathcal{K}$ is the set of objective function indices (e.g. $\mathcal{K} = \{1, 2, \ldots, K\}$). The task on multi-factorial optimization problems is to acquire K $(= |\mathcal{K}|)$ optimal solutions $x_k^* = \arg\min_{x \in \mathcal{X}} f_k(x)$ $(k \in \mathcal{K})$.

Multiple objective functions f_k $(k \in \mathcal{K})$ exist but design variable x is common among them. Unlike multi-objective optimization to search for the optimal trade-off, the Pareto front, among objective functions, multi-factorial optimization searches K optimal solutions on K objective functions.

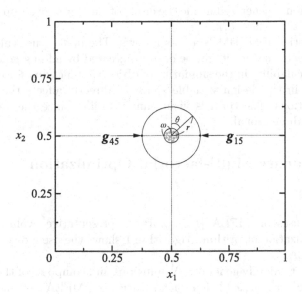

Fig. 1. Design variable space $\mathcal{X}$ of MFDMP with goal points g_{15} and g_{45}

2.2 Multi-factorial Distance Minimization Problem

In this work, we use multi-factorial distance minimization problem (MFDMP) [7], which is a benchmark problem framework for multi-factorial optimization. MFDMP is an extension of multi-objective distance minimization problem [8,9] frequently used in multi-objective optimization benchmarking.

As shown in Fig. 1, MFDMP has a $d = 2$ dimensional design variable space $\mathcal{X} = [0, 1]^2$. g_k is a goal point indicating the optimal point x_k^* of objective function f_k. Goal point g_k is set on the circle of the center position $c = (0.5, 0.5)$ and the radius r. The maximum number of goal positions is limited to 60 in this work. We treat the circle with the radius r as an analog clock and set goal points on minute positions. The clock-wise angle θ based on the center position $c = (0.5, 0.5)$ and $x_1 = 0.5$ is used to set each goal point g_k. Goal point g_k at k minute position of the analog clock is given by

$$g_k = \left(r \cdot \sin\left(k \cdot \left(\frac{2\pi}{60} \right) \right) + 0.5, \; r \cdot \cos\left(k \cdot \left(\frac{2\pi}{60} \right) \right) + 0.5 \right). \tag{1}$$

The minimizing objective function f_k corresponding to the goal point g_k is given by

$$f_k(x) = ||x - g_k||, \tag{2}$$

where $||\cdot||$ is the Euclidean distance. Figure 1 shows a case with the goal point $\boldsymbol{g}_{15}$ at 15 min position and the goal point $\boldsymbol{g}_{45}$ at 45 min position on the analog clock circle. We minimize two objective functions $f_{15}(\boldsymbol{x})$ and $f_{45}(\boldsymbol{x})$ in this case. The initial population is generated inside the circle of the center position $\boldsymbol{c} = (0.5, 0.5)$ and the radius ω in this work.

Characteristics of MFDMPs are as follows. The first is the scalability in the number of objectives since it can be easily increased by adding goal points. The second is the scalability in the similarity of objective functions since the distance of goal points in the design variable space $\mathcal{X}$ directly affects the similarity of objective functions. The third is visual analyzability since the design variable space is $d = 2$ dimensional.

3   Evolutionary Multi-factorial Optimization

3.1   Method

This work focuses on MFEA [1, 2] as the representative evolutionary multi-factorial optimization algorithm. Algorithm 1 shows the pseudo-code of the conventional MFEA.

MFEA first randomly generates N solutions and composes of the parent population $\mathcal{P} = \{\boldsymbol{x}^1, \boldsymbol{x}^2, \ldots, \boldsymbol{x}^N\}$. For each solution $\boldsymbol{x}^i$, MFEA assigns a skill factor s^i indicating target objective function f_{s^i} and composes of the skill factor set $\mathcal{S} = \{s^1, s^2, \ldots, s^N\}$. MFEA evaluates each solution $\boldsymbol{x}^i$ on the target objective function f_{s^i} specified by its skill factor s^i. To generate new solutions, MFEA randomly choose two parents $\boldsymbol{x}^p$ and $\boldsymbol{x}^q$ from the parent population $\mathcal{P}$. MFEA then applies a crossover to them when the two parents $\boldsymbol{x}^p$ and $\boldsymbol{x}^q$ have the same skill factor (i.e. $s^p = s^q$), or under the probability rmp even if they do not have the same skill factor (i.e. $s^p \neq s^q$). A mutation to perturb design variable values is also applied to the obtained offspring $\boldsymbol{x}^{p'}$ and $\boldsymbol{x}^{q'}$, and they are added to the offspring population $\mathcal{P}'$. Skill factors $s^{p'}$ and $s^{q'}$ of offspring are inherited from the skill factors s^p and s^q of their parents. The above offspring generation is repeated until the size of the offspring population $\mathcal{S}'$ reaches to N, which is the same size of the parent population $\mathcal{P}$. For each solution in the combined population $\mathcal{P} \cup \mathcal{P}'$, we calculate scalar fitness, which is the rank value on each targeting objective function specified by skill factor. MFEA then selects the best N solutions from the combined population $\mathcal{P} \cup \mathcal{P}'$ as the new parent population $\mathcal{P}$ in terms of scalar fitness.

MFEA repeats the above process and simultaneously optimizes multiple objective functions with the single population $\mathcal{P}$.

3.2   Issue Focus

The crossover to recombine solutions with different skill factors has an important role in evolutionary multi-factorial optimization since it enhances the simultaneous optimization of multiple objective functions. Therefore, the selection of

---

Algorithm 1. Conventional MFEA [1,2]

1: $\mathcal{P} = \{x^1, x^2, \ldots, x^N\} \leftarrow$ Randomly generate the population   ▷ Parent population
2: $\mathcal{S} = \{s^1, s^2, \ldots, s^N\} \leftarrow$ Evenly assign skill factors   ▷ <u>Skill factors of $\mathcal{P}$</u>
3: Evaluate each $x^i \in \mathcal{P}$ on objective function f_{s^i}
4: **for** $g \leftarrow 1, 2, \ldots, G$ **do**
5:      $\mathcal{P}' \leftarrow \emptyset$ ▷ Offspring population
6:      $\mathcal{S}' \leftarrow \emptyset$ ▷ <u>Skill factors of $\mathcal{P}'$</u>
7:      **while** $|\mathcal{P}'| < N$ **do**
8:          $p, q \leftarrow$ Randomly choose $(1, 2, \ldots, N)$
9:          **if** $s^p = s^q \vee \text{rand}(0,1) < rmp$ **then**
10:             $x^{p'}, x^{q'} \leftarrow$ Crossover (x^p, x^q)
11:         **else**
12:             $x^{p'}, x^{q'} \leftarrow$ Copy (x^p, x^q)
13:         **end if**
14:         $\mathcal{P}' \leftarrow \mathcal{P}' \cup \{x^{p'}, x^{q'}\} =$ Mutation $(x^{p'}, x^{q'})$
15:         $\mathcal{S}' \leftarrow \mathcal{S}' \cup \{s^{p'}, s^{q'}\} =$ Inherit (s^p, s^q)
16:     **end while**
17:     Evaluate each $x^{i'} \in \mathcal{P}'$ on objective function $f_{s^{i'}}$
18:     Calculate scalar fitness $(\mathcal{P} \cup \mathcal{P}')$
19:     $\mathcal{P}, \mathcal{S} \leftarrow$ Select best N solutions $(\mathcal{P} \cup \mathcal{P}', \mathcal{S} \cup \mathcal{S}')$
20: **end for**
21: **return** $x_k^* = \arg\min_{x \in \mathcal{P}} f_k(x)$ $(k \in \mathcal{K})$

---

parents to be crossed strongly affects crossover effects. However, the conventional MFEA randomly chooses parents without considering similarities among objective functions. As a result, solutions with skill factors of dissimilar objective functions become pairs of parents and are crossed. This would be a barrier to evolutionary multi-factorial optimization.

Evolutionary multi-factorial optimization would be further enhanced if we could estimate similarities among objective functions, encourage crossovers of solutions with similar skill factors and suppress crossovers of solutions with dissimilar skill factors.

4 Proposal: MFEA Using Objective Similarity Based Parent Selection

4.1 Summary

In this work, we propose an MFEA-based algorithm that estimates similarities of objective functions and reflects them to parent selection probabilities to enhance evolutionary multi-factorial optimization. The proposed method encourages crossovers of solutions with skill factors of similar objective functions and suppresses crossovers of solutions with skill factors of dissimilar objective functions. Algorithm 2 shows the pseudo-code of the proposed algorithm. Differences from the conventional MFEA [1,2] is highlighted with blue.

Algorithm 2. Proposed MFEA

1: $\mathcal{P} = \{x^1, x^2, \ldots, x^N\} \leftarrow$ Randomly generate the population $\quad \triangleright$ Parent population
2: $\mathcal{S} = \{s^1, s^2, \ldots, s^N\} \leftarrow$ Evenly assign skill factors $\quad\quad\quad\quad \triangleright$ Skill factors of $\mathcal{P}$
3: Evaluate each $x^i \in \mathcal{P}$ on objective function f_{s^i}
4: **for** $g \leftarrow 1, 2, \ldots, G$ **do**
5: $\quad$ **for each** $k \in \mathcal{K}$ **do**
6: $\quad\quad m_k \leftarrow \left(\frac{\sum_{x \in \mathcal{P}_k} x_1}{|\mathcal{P}_k|}, \frac{\sum_{x \in \mathcal{P}_k} x_2}{|\mathcal{P}_k|}, \ldots, \frac{\sum_{x \in \mathcal{P}_k} x_d}{|\mathcal{P}_k|} \right) \quad \triangleright$ Mean position, Eq. (3)
7: $\quad\quad v_k \leftarrow m_k - c \quad\quad\quad\quad\quad\quad\quad\quad\quad\quad\quad \triangleright$ Search direction, Eq. (4)
8: $\quad$ **end for**
9: $\quad$ **for each** $k \in \mathcal{K}$ **do**
10: $\quad\quad$ **for each** $l \in \mathcal{K}$ **do**
11: $\quad\quad\quad S_{k,l} \leftarrow \frac{\cos(v_k, v_l) + 1}{2} \quad\quad\quad\quad\quad\quad \triangleright$ Objective similarity, Eq. (5)
12: $\quad\quad\quad P_{k,l} \leftarrow \frac{S_{k,l}^{\alpha}}{\sum_{m \in \mathcal{K}} S_{k,m}^{\alpha}} \quad\quad\quad\quad\quad \triangleright$ Selection probability, Eq. (6)
13: $\quad\quad$ **end for**
14: $\quad$ **end for**
15: $\quad \mathcal{P}' \leftarrow \emptyset \quad\quad\quad\quad\quad\quad\quad\quad\quad\quad\quad\quad\quad \triangleright$ Offspring population
16: $\quad \mathcal{S}' \leftarrow \emptyset \quad\quad\quad\quad\quad\quad\quad\quad\quad\quad\quad\quad\quad \triangleright$ Skill factors of $\mathcal{P}'$
17: $\quad$ **while** $|\mathcal{P}'| < N$ **do**
18: $\quad\quad k \leftarrow$ Randomly choose a skill factor $(\mathcal{K})$
19: $\quad\quad x^p \leftarrow$ Randomly choose first parent $(\mathcal{P}_k)$
20: $\quad\quad l \leftarrow$ Probabilistically choose a skill factor $(\mathcal{K}, P_{k,1}, P_{k,2}, \ldots, P_{k,K})$
21: $\quad\quad x^q \leftarrow$ Randomly choose second parent $(\mathcal{P}_l)$
22: $\quad\quad x^{p'}, x^{q'} \leftarrow$ Crossover (x^p, x^q)
23: $\quad\quad \mathcal{P}' \leftarrow \mathcal{P}' \cup \{x^{p'}, x^{q'}\} =$ Mutation $(x^{p'}, x^{q'})$
24: $\quad\quad \mathcal{S}' \leftarrow \mathcal{S}' \cup \{s^{p'}, s^{q'}\} =$ Inherit (s^p, s^q)
25: $\quad$ **end while**
26: $\quad$ Evaluate each $x^{i'} \in \mathcal{P}'$ on objective function $f_{s^{i'}}$
27: $\quad$ Calculate scalar fitness $(\mathcal{P} \cup \mathcal{P}')$
28: $\quad \mathcal{P}, \mathcal{S} \leftarrow$ Select best N solutions $(\mathcal{P} \cup \mathcal{P}', \mathcal{S} \cup \mathcal{S}')$
29: **end for**
30: **return** $x_k^* = \arg\min_{x \in \mathcal{P}} f_k(x) \ (k \in \mathcal{K})$

4.2 Search Direction

For each skill factor k indicating target objective function f_k $(k \in \mathcal{K})$, the proposed algorithm estimates the search direction v_k in the design variable space $\mathcal{X}$. 5–8 lines of Algorithm 2 corresponds to this process.

As with the conventional MFEA, the proposed algorithm also assigns a skill factor s^i to each solution x^i in the parent population $\mathcal{P}$. That is, solution x^i is evaluated and responsible for objective function f_{s^i}. For each skill factor $k \in \mathcal{K}$, we calculate the mean position vector m_k of $\mathcal{P}_k = \{x^i \in \mathcal{P} | s^i = k\}$, which is the solution set with skill factor k in the parent population $\mathcal{P}$. The mean position vector m_k of skill factor k is given by

$$ m_k = \left(\frac{\sum_{x \in \mathcal{P}_k} x_1}{|\mathcal{P}_k|}, \frac{\sum_{x \in \mathcal{P}_k} x_2}{|\mathcal{P}_k|}, \ldots, \frac{\sum_{x \in \mathcal{P}_k} x_d}{|\mathcal{P}_k|} \right), \quad\quad (3) $$

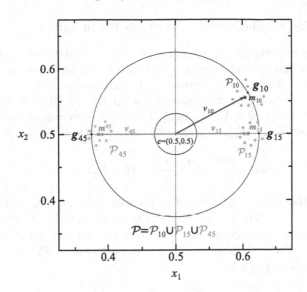

Fig. 2. Three search directions for three objective functions, goal points (Color figure online)

where $\mathcal{P}_k \subseteq \mathcal{P}$ and $\mathcal{P} = \bigcup_{k \in \mathcal{K}} \mathcal{P}_k$.

Next, for each skill factor $k \in \mathcal{K}$, we calculate the search direction vector v_k in the design variable space $\mathcal{X}$. The search direction vector v_k is calculated with the central position c of the design variable space $\mathcal{X}$ by

$$v_k = m_k - c. \tag{4}$$

In the case of MFDMP, we use $c = (0.5, 0.5)$. c indicates the mean position of the initial solutions. In this way, each skill factor k is characterised by the search direction v_k from the central position c to the mean position m_k of the current solutions $\mathcal{P}_k$ with skill factor k in the design variable space $\mathcal{X}$.

Figure 2 shows an example. The central position $c = (0.5, 0.5)$ is shown as the blue point. There are three goals g_{10}, g_{15}, and g_{45}, which respectively correspond to objective functions f_{10}, f_{15}, and f_{45}. Three subset populations $\mathcal{P}_{10}, \mathcal{P}_{15}$, and $\mathcal{P}_{45}$ with skill factors 10, 15, and 45 are shown with different colored circles. Also, the mean position vectors m_{10}, m_{15}, and m_{45} are shown as rectangles, and the search direction vectors v_{10}, v_{15}, and v_{45} are also depicted. Note that the three goal positions g_{10}, g_{15}, and g_{45} are unknown for optimizer. However, we see the similarity of f_{10} and f_{15} from the their search directions v_{10} and v_{15}. Also, we see the dissimilarity of f_{15} and f_{45} from their search directions v_{15} and v_{45}.

4.3   Objective Similarity and Selection Probability

Next, the proposed algorithm estimates objective similarities among objective functions and calculates selection probabilities based on the objective similarities. 9–14 lines of Algorithm 2 correspond to this process.

The proposed algorithm quantitatively estimates objective similarities among objective functions f_k ($k \in \mathcal{K}$) by using the search direction vectors v_k ($k \in \mathcal{K}$). The objective similarity $S_{k,l}$ between objective functions f_k and f_l is given by

$$S_{k,l} = \frac{\cos(v_k, v_l) + 1}{2} \quad (k \in \mathcal{K}, l \in \mathcal{K}). \tag{5}$$

Thus, we use the cosine similarity of the search directions v_k and v_l. $S_{k,l}$ is in the value range $[0, 1]$. The higher $S_{k,l}$, the more similar objective functions.

We then calculate selection probabilities based on the objective similarities obtained above. For skill factor k ($\in \mathcal{K}$), the probabilities $P_{k,l}$ ($l \in \mathcal{K}$) to select the skill factor l are given by

$$P_{k,l} = \frac{S_{k,l}^{\alpha}}{\sum_{m \in \mathcal{K}} S_{k,m}^{\alpha}} \quad (k \in \mathcal{K}, l \in \mathcal{K}), \tag{6}$$

where $\alpha = [0, \infty]$ is the exponent parameter to bias the selection probability. $P_{k,l}$ is in the value range $[0, 1]$, and $\sum_{l \in \mathcal{K}} P_{k,l} = 1$. $\alpha = 0$ indicates that the selection probabilities of any skill factors are equal and is equivalent to the random selection. The probability of selecting solutions with similar objective functions increases as the increase of α.

4.4   Parent Selection

The proposed algorithm selects pairs of parents based on the selection probabilities obtained above. 18–21 lines of Algorithm 2 correspond to this process.

To select a pair of parents x^p and x^q from the parent population $\mathcal{P}$, we randomly choose one skill factor k among the set of skill factor indices $\mathcal{K}$. We then randomly choose the first parent x^p among $\mathcal{P}_k = \{x^i \in \mathcal{P} | s^i = k\}$, which is the solution set with skill factor k in the parent population $\mathcal{P}$. Next, we select the skill factor l of the second parent solution based on the probabilities $P_{k,j}$ ($j \in \mathcal{K}$). For the selected skill factor l, we randomly choose the second parent x^q among $\mathcal{P}_l = \{x^i \in \mathcal{P} | s^i = l\}$, which is the solution set with skill factor l in the parent population $\mathcal{P}$.

In this way, the proposed algorithm makes pairs of parents according to the selection probabilities based on the objective similarities. In contrast, the conventional MFEA makes pairs of parents randomly without considering the objective similarities. The proposed selection encourages crossovers of parents with skill factors of similar objective functions and suppresses crossovers of parents with skill factors of dissimilar objective functions.

Table 1. MFDMPs used in this paper

Problem name	Number of objectives $K(=|\mathcal{K}|)$	Goal positions (objective indices) $\mathcal{K}$
Uniform MFDMP-4	4	{0, 15, 30, 45}
Uniform MFDMP-10	10	{0, 6, 12, 18, 24, 30, 36, 42, 48, 54}
Uniform MFDMP-12	12	{0, 5, 10, 15, 20, 25, 30, 35, 40, 45, 50, 55}
Biased MFDMP-A	12	{0, 30, 31, 32, 33, 34, 35, 36, 37, 38, 39, 40}
Biased MFDMP-B	12	{0, 1, 2, 3, 4, 5, 30, 31, 32, 33, 34, 35}
Biased MFDMP-C	12	{0, 1, 2, 15, 16, 17, 30, 31, 32, 45, 46, 47}

5   Experimental Setting

5.1   Algorithm

We compare the search performances of three algorithms: (i) the conventional MFEA [1,2], (ii) the conventional single-objective EA (SOEA), and (iii) the proposed MFEA using the objective similarity based parent selection.

For all algorithms, we use the SBX crossover (the distribution index $\eta_c = 2$ and the crossover ratio 1.0) and the polynomial mutation (the distribution index $\eta_m = 5$ and the mutation rate 0.5). The population size is set to 100, and the total number of generations is set to $G = 80$. Each algorithm is executed 100 times, and we compare the average result.

For the conventional MFEA, rmp is set to 0.3. The conventional SOEA also uses the single population and assigns a skill factor to each solution. However, any interactions among solutions with different skill factors are not performed. That is, a solution for a skill factor is always crossed with a solution with the same skill factor.

5.2   Metric

As the metric to evaluate the multi-factorial optimization performance, we use D metric [7] given by

$$D = \frac{1}{|\mathcal{K}|} \sum_{k \in \mathcal{K}} \min_{x \in \mathcal{P}} f_k(x). \tag{7}$$

D is the average objective function value of the best objective function values on $K(= |\mathcal{K}|)$ objective functions in the population $\mathcal{P}$. The smaller D, the better multi-factorial optimization performance.

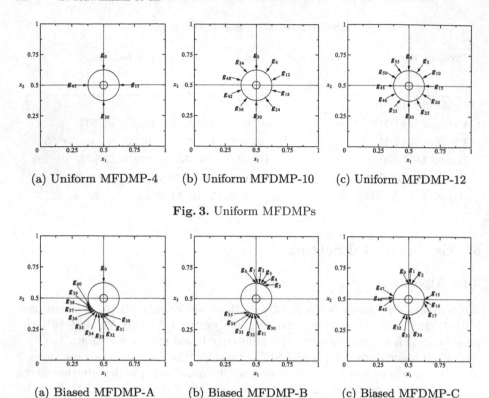

(a) Uniform MFDMP-4   (b) Uniform MFDMP-10   (c) Uniform MFDMP-12

Fig. 3. Uniform MFDMPs

(a) Biased MFDMP-A   (b) Biased MFDMP-B   (c) Biased MFDMP-C

Fig. 4. Biased MFDMPs

5.3   Problems

For MFDMPs, we use the radius $r = 0.125$ to set goal positions on the circle. Also, we use the radius $\omega = 0.03125$ to set the circle for the solution initialization.

Table 1 shows settings of six MFDMPs used in this work. As mentioned before, the goal-setting circle is equally divided into 60 intervals, such as minute intervals of the analog clock. The first three MFDMPs are uniform MFDMPs, in which the goal positions are uniformly distributed on minute positions. We use the uniform MFDMP-4, -10, and -12 with $K = \{4, 10, 12\}$ goal points, respectively. Goal positions are visually shown in Fig. 3. The other three MFDMPs are biased MFDMPs, in which the goal positions are biasedly distributed on minute positions. We use the biased MFDMP-A, -B, and -C with $K = 12$ goal points. Goal positions are visually shown in Fig. 4. In the biased MFDMP-A, one goal is isolatedly positioned from others. The biased MFDMP-B has two groups of goals, and the two groups are positioned on the opposite side. Each group has six goals, and they are densely distributed. The biased MFDMP-C has four groups of goals, and the four groups are uniformly positioned. Each group has three goals, and they are densely distributed.

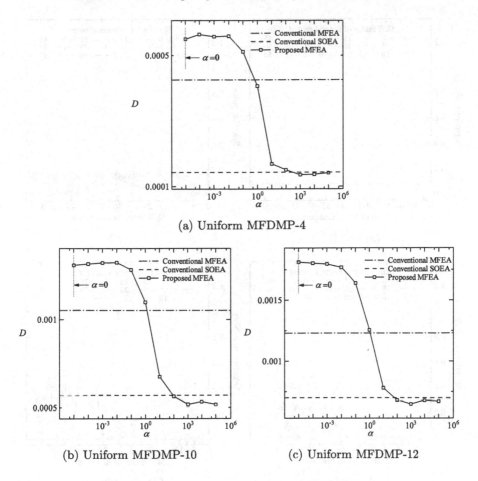

(a) Uniform MFDMP-4

(b) Uniform MFDMP-10   (c) Uniform MFDMP-12

Fig. 5. Results of D on uniform MFDMPs

6 Experimental Results and Discussion

6.1 Results on Uniform MFDMPs

Figure 5 shows results of the proposed algorithm on uniform MFDMPs when we vary the parameter α. The increase of α emphasizes the bias of the selection probabilities based on objective similarities. Note that the horizontal axis is the logarithmic scale, and results with $\alpha = 0$ are exceptionally plotted on $\alpha = 10^{-5}$. Each figure also involves results of the conventional SOEA and MFEA as horizontal lines.

$\alpha = 0$ indicates the random selection without considering the objective similarities. We see that D decreases as α increases from $\alpha = 0$. That is, the selection bias based on objective similarities improves the multi-factorial optimization performance. For three uniform MFDMPs, $\alpha = 10^3$ achieves the best D values, which are smaller than ones of the conventional SOEA and MFEA.

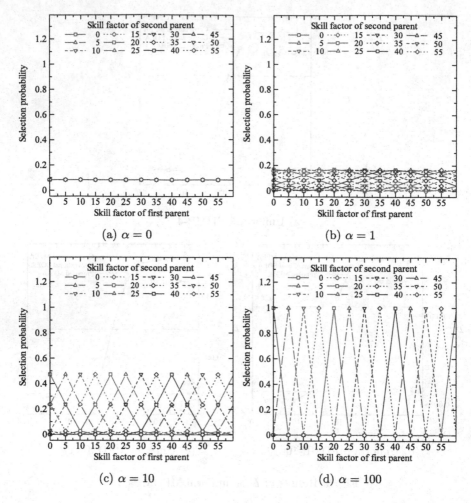

Fig. 6. Selection probabilities on the uniform MFDMP-K12 (Color figure online)

Since we limited the initialization area of the population, the cooperative search among solutions with different skill factors does not work well on the conventional MFEA. As a result, the conventional MFEA is worse than the conventional SOEA. Even in this problem situation, the cooperative search on the proposed algorithm works well, and the proposed algorithm achieves the best performance by controlling the solution selection bias.

Next, we focus only on the uniform MFDMP-12 and observe the selection probabilities obtained by the proposed algorithm. Figure 6 shows the selection probabilities when we vary the parameter α. The horizontal axis indicates the skill factors of the first parent. Each figure has twelve plots, which are skill factors of second parents. That is, for each skill factor of the first parent on

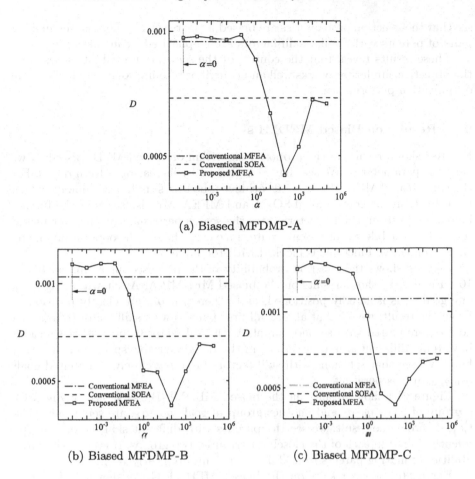

(a) Biased MFDMP-A

(b) Biased MFDMP-B

(c) Biased MFDMP-C

Fig. 7. Results of D on biased MFDMPs

the horizontal axis, we see that the selection probabilities of skill factors of the second parent.

In Fig. 6(a) with $\alpha = 0$, we see that the selection probabilities for all objective functions are flat, which brings the random parent selection without considering the objective similarities. In Fig. 6(b) with $\alpha = 1$, we see a selection bias based on the objective similarities. Note that the horizontal axis indicates skill factor indices from 0 to 59, which are minute positions of the analog clock shown before. Here, we focus on the solid red line with the rectangle marker showing the selection probabilities of the second parent with the skill factor 0. The highest selection probability can be seen when the first parent is also with the skill factor 0. The second highest probability can be seen when the first parent is with the skill factor 5 or 55, which is the neighborhood of the skill factor 0. Thus, the selection probabilities of the second parent with the skill factor 0 decrease as increasing the distance from the skill factor 0. From Fig. 6(c) and Fig. 6(d), we see that the selection probabilities are biased by increasing α. That is, the making pairs of parents with similar skill factors are emphasized by increasing α.

These results reveal that the control of the selection probabilities based on the objective similarities works well and contributes to improving multi-factorial optimization performance.

6.2   Results on Biased MFDMPs

Figure 7 shows results of the proposed algorithm on biased MFDMPs when we vary the parameter α. We see that D decreases by increasing α from $\alpha = 0$. For the three biased MFDMPs, $\alpha = 10^2$ achieves the best search performance, which is better than the conventional SOEA and MFEA. We also see that the further increase of α from $\alpha = 10^2$ deteriorates the search performance. The over-biased selection probability is not appropriate since even the search cooperation among similar objective functions is blocked with too large α.

Figure 8 shows the selection probability of the proposed algorithm with $\alpha = 10^2$. Figure 8(a) shows results on the biased MFDMP-A. As shown in Fig. 4(a), one goal g_0 is isolatedly positioned, and others g_{30}–g_{40} are closely positioned. From the result, we see that almost all first parents with skill factor 0 targeting goal g_0 are paired with second parents with skill factor 0 since other solutions have quite different search directions in the design variable space. On the other hand, we see that solutions with skill factors targeting closely distributed goals g_{30}–g_{40} are cooperating.

Figure 8(b) shows results on the biased MFDMP-B. As shown in Fig. 4(b), a group of goals g_0–g_5 and another group of goals g_{30}–g_{35} are separately positioned. From the result, we see the parents with different skill factors are frequently paired in each of the closely distributed two groups. However, the probabilities of mating parents from different groups are nearly zero.

Figure 8(c) shows results on the biased MFDMP-C. As shown in Fig. 4(c), there are four groups of goals. The first one involves g_0–g_2, the second one involves g_{15}–g_{17}, the third one involves g_{30}–g_{32}, and the fourth one involves g_{45}–g_{47}. From the result, we see that the selection probabilities are shared in each group. In other words, the solution resources are shared in each group even if their skill factors are different.

These results verified that the proposed method encouraged pairing for solutions with similar objective functions and suppressed pairing for solutions with dissimilar objective functions. The results also clarified that the proposed similarity-based parent selection improved the multi-factorial optimization performance.

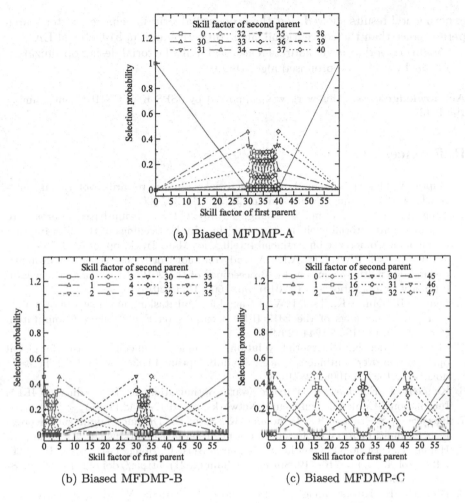

(a) Biased MFDMP-A

(b) Biased MFDMP-B   (c) Biased MFDMP-C

Fig. 8. Selection probabilities on the biased MFDMPs

7 Conclusions

To accelerate the evolutionary multi-factorial optimization by encouraging crossovers of solutions with similar target objective functions and suppressing crossovers of solutions with dissimilar target objective functions, in this work, we proposed an evolutionary algorithm estimating the similarities of objective functions and utilizing them for the parent selection determining pairs of solutions to be crossed. We used the multi-factorial distance minimization problems to verify the effectiveness of the proposed algorithm. Experimental results showed that the distance relations among objective functions are matched to the similarities among objective functions estimated by the proposed method. Also, experimental results showed that the proposed algorithm achieves better search performance than the conventional single-objective and multi-factorial EAs.

As future work, we will address real-world multi-factorial design optimization problems by using the proposed algorithm.

Acknowledgments. This work was supported by JSPS KAKENHI Grant Number 19K12135.

References

1. Gupta, A., Ong, Y.S., Feng, L.: Multifactorial evolution: towards evolutionary multitasking. IEEE Trans. Evol. Comput. **20**(3), 343–357 (2015)
2. Tang, J., Chen, Y., Deng, Z., Xiang, Y., Joy, C.P.: A group-based approach to improve multifactorial evolutionary algorithm. In: Proceedings of the 27th International Joint Conference on Artificial Intelligence Main Track, pp. 3870–3876 (2018)
3. Jin, C.C., Tsai, P.W., Qin, A.K.: A study on knowledge reuse strategies in multitasking differential evolution. In: Proceedings of the 2019 IEEE Congress on Evolutionary Computation (CEC 2019), pp. 1564–1571 (2019)
4. Song, D.H., Qin, A.K., Tsai, P.W., Liang, J.J.: Multitasking multi-swarm optimization. In: Proceedings of the 2019 IEEE Congress on Evolutionary Computation (CEC 2019), pp. 1937–1944 (2019)
5. Storn, R., Price, K.: Differential evolution - a simple and efficient heuristic for global optimization over continuous spaces. J. Glob. Optim. **11**, 341–359 (1997). https://doi.org/10.1023/A:1008202821328
6. Kennedy, J., Eberhart, R.: Particle swarm optimization. In: Proceedings of IEEE International Conference on Neural Networks, pp. 1942–1948 (1995)
7. Kawakami, S., Takagi, T., Takadama, K., Sato, H.: Distance minimization problems for multi-factorial evolutionary optimization benchmarking. In: Abraham, A., Hanne, T., Castillo, O., Gandhi, N., Nogueira Rios, T., Hong, T.-P. (eds.) HIS 2020. AISC, vol. 1375, pp. 710–719. Springer, Cham (2021). https://doi.org/10.1007/978-3-030-73050-5_69
8. Ishibuchi, H., Hitotsuyanagi, Y., Tsukamoto, N., Nojima, Y.: Many-objective test problems to visually examine the behavior of multiobjective evolution in a decision space. In: Schaefer, R., Cotta, C., Kołodziej, J., Rudolph, G. (eds.) PPSN 2010. LNCS, vol. 6239, pp. 91–100. Springer, Heidelberg (2010). https://doi.org/10.1007/978-3-642-15871-1_10
9. Ishibuchi, H., Yamane, M., Akedo, N., Nojima, Y.: Many-objective and many-variable test problems for visual examination of multiobjective search. In: Proceedings of the 2013 IEEE Congress on Evolutionary Computation (CEC 2013), pp. 1491–1498 (2013)

Aural Language Translation with Augmented Reality Glasses

Ian N. Hovde, Forrest S. Kelley, Ryan J. Kearney, and Douglas E. Dow[⊠]

Electrical and Computer Engineering, School of Engineering, Wentworth Institute of
Technology, Boston, MA 02115, USA
dowd@wit.edu

Abstract. Communication is core to human activity. Communication across language barriers is necessary for many types of travel, business, diplomacy, environmental or society movements, and friendships. Dictionaries and software tools help with translation, but often are inadequate for in-person practical communication. The natural method is to use a human translator, but that requires sufficient skill in both languages and of the content area being discussed. The human translator as a third party diminishes privacy. A software based mobile system could potentially improve privacy, availability, and knowledge of the content area. Such a system could be a wearable mobile device connected with web services. This project developed and tested an aural translation system using augmented reality (AR) glasses with audio capabilities connected with a smartphone and Amazon Web Services (AWS) for transcription, translation, and conversion of text back to audio. A prototype was developed based on a Bose AR sunglasses system, and was tested for phrases in English, Spanish and German. The results had reasonable accuracy and processing times. Further development and testing are necessary for wide application, but the results support further development.

Keywords: Cloud services · Transcription · German · Spanish · English · Augmented Reality. AR

1 Introduction

Communication is one of the most essential parts of society. It is not only necessary in everyday life, but in travel, finance and politics. The variety of verbal languages used for social interaction and communication globally makes it difficult to communicate in many situations. There are so many languages across the world that learning them all would be impossible. However, business, culture, and education drive people to try and communicate across different languages. Many techniques and tools are available to help translation, but none are fully adequate. A primary hindrance is how intrusive or distracting using the tool is in a physically present person-to-person communication.

The translation method based on natural processes uses a human translator to assist in the translation. The human translator listens to the sentences one person speaks in

T. Nakano (Ed.): BICT 2021, LNICST 403, pp. 61–70, 2021.
https://doi.org/10.1007/978-3-030-92163-7_6

the first language, understands the meaning, translates the meaning to the second language, and speaks the new translated sentences aloud for the second person to hear in the second language. For this natural process to work, the human translator needs to be able understand both languages, including the content. The content often includes jargon, specialized words and implied meanings based on context. Finding a suitable and available translator is often difficult. Not only does the translator have an undue influence in the conversation by the choice and tone of phrases, but also gains an in-depth knowledge of the conversation. Having a third person with influence and knowledge of a conversation may be concerning in negotiations involving politics, business or romance.

Considering the advantages and disadvantages of the natural process of a human translator, an engineered system might be able to alleviate some of the disadvantages while maintaining some benefits. An impersonal tool that would automate and individualize language translation would give power back to people who need to communicate across language barriers. A tool based on discrete mobile microphones and speakers, and software transcription and translation based on web services might be able to lower the price, increase access and have context expertise (if the utilized web services include that context linguistic knowledge).

Several manual or software-based methods for translation exist or are being developed. Traditional or bilingual dictionaries are the most basic tool for language translation. Going from definition to definition can allow for a confident word translation and the elimination of a third person in the conversation. However, using a dictionary for each word would be time consuming, and sometimes leads to poor understanding of grammar and sentence structure. Certain jargon, slang, grammar, and idioms can be improperly translated by standard dictionaries. Such an improperly translated word or phrase may change the entire meaning of a statement, leaving both parties confused, going forward with misunderstanding. Using a dictionary during in-person communication is not a viable solution.

Besides a physical or electronic dictionary, a web service such as Google Translate (Alphabet, Mountain View, CA; https://translate.google.com/) can make translations of text sentences or paragraphs. Nevertheless, such web services have not been routinely utilized for in-person conversations due to the hassle and disruption to the flow of conversation. Google Translate analyzes input text, translates and displays the text back for the user to read. Google translate works relatively well for both singular words and full sentences. As a result, Google Translate is a popular secondary tool for learning a language [6]. For use in real time conversations, Google Translate has drawbacks. With no built-in audio feature and transcription to text, the user would be required to attempt to enter text for the verbal discussion. This would be difficult and not practical. It would also be a burden to enter in long sentences by phonics only in a language that the user does not know.

An example of audio input and output for Google Translate is available with the Google Pixel 2 earbuds [9]. Google's earbuds connect with the Google Pixel mobile phone, providing a translation pathway from spoken audio in one language. They have the added advantage by Google Translate and text-speech to play through earbuds, which allows for almost "real-time" translation to the user [1]. This audio-to-audio translation functionality is good but is currently available only on a Google Pixel phone, reducing accessibility.

Another product that could help cross the language barrier of communication is the Vuzix Augmented Reality Glasses [7]. The system inputs verbally spoken audio, translates and displays the translated text on AR glasses display for the user to read [7]. This works well for translation, but the user would have to read display text each time, which may be distracting during a conversation.

A translation system that goes from verbal audio in one language to verbal audio in another language using web-based tools would be beneficial. Some models of AR glasses have capabilities beyond visual to include audio, with built in microphones and speakers. Figure 1 shows an illustration of such an AR glasses based aural translation system. The goal of this project was to develop and test the use of AR audio glasses with mobile device software communicating to web services for translation between languages.

Fig. 1. Illustration of 2 people needing to communicate and using the proposed AR audio classes with web services for translation.

2   Design

Bose (Framingham, Massachusetts) has developed prototypes of AR Sunglasses that have microphones and speakers to record and play audio back to the user. Prototypes of these AR audio glasses were used for the prototype of this project. Using the Bose AR Glasses and commercially obtained mobile phone, the primary development work for this project was done with software development.

The envisioned design involved an app that ran on the user's smartphone, and connected with the Bose AR sunglasses. Amazon Web Services (AWS, Seattle, WA) was utilized for the transcription and translation services.

The user would use built in AR features of the glasses as the human computer interface (HCI) to signal that the system should start recording the speech and initialize the translation procedure. For the prototype, the application was started when the user double tapped on the side of the glasses. They heard a small acknowledgement ping that their action was successful and that the program was ready to record. Then whatever was spoken or heard was recorded. When the verbal sentence or phrase was done the user would again double tap to let the system know the recording should stop. The application would receive the recording as an audio file via Bluetooth from the AR glasses. The app would send the audio file to AWS to transcribe the audio to text. The transcribed text would then be translated to the desired language using AWS. The translated text would then be converted back to audio and then played to the user through the built-in speaker on the glasses. The whole process from the AR audio glasses through the smartphone to AWS is shown in Fig. 2.

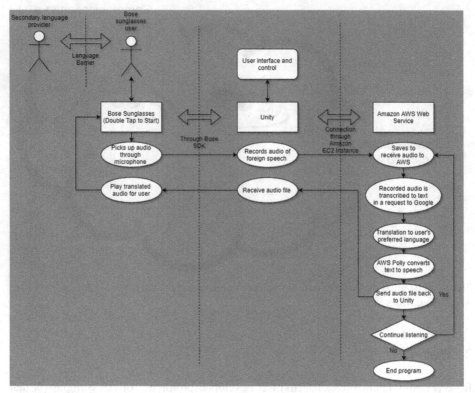

Fig. 2. Block diagram of audio file path. Shows the full cycle the application goes through to provide the user an accurate translation.

Several software modules were developed for the prototype. The modules included the code that ran on the mobile phone and the code that ran on the server. The code running on the phone handled all the communication with the glasses and also included the user interface. The server code handled the transcription and translation. The communication between these two modules occurred with HTTP Restful API requests. The overall workflow of this design is illustrated in Fig. 3.

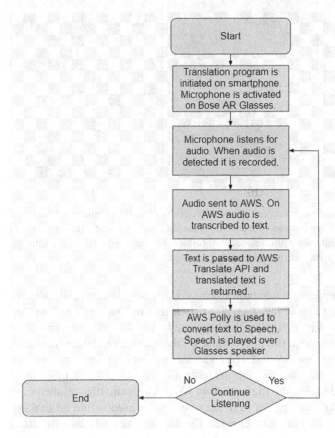

Fig. 3. Flow chart highlights the full cycle of the program. This shows steps of translation through the Amazon Web Services (AWS) for transcription and translation.

The rationale for this design includes the available technologies contained in the Bose glasses. These features included a built-in microphone and a pair of Bose quality speakers [4]. It also had AR touch and gesture features, which were used in the proposed solution. This allowed for several gestures to facilitate the HCI, allowing the user to keep their phone in their pocket, and their focus on who they are talking with. The front-end app allowed for a visual interface the user could navigate the system with. While it is not the primary method of navigation it helped the user to comprehend the product features. Examples of the mobile interface are shown in Fig. 4 and follow a path of connecting to the glasses over Bluetooth, bringing the user to the home screen, then opening up paths for settings, language alterations, and a help page.

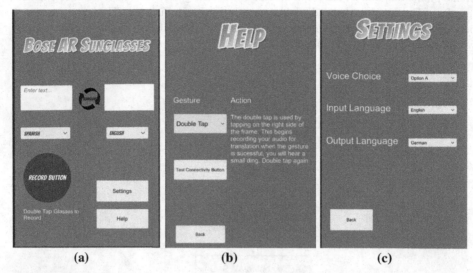

(a)   **(b)**   **(c)**

Fig. 4. Unity Interface. (A) is the main screen of the application. Here the language settings can be set and the features of the translator can be used. (B) is the help page. On this page you can see if the glasses are connected and information about AR gestures are displayed. (C) is the settings pages where you can set additional settings like voice preferences.

The front-end for the application was built in Unity. Unity is a platform that supports AR game development [2]. The user interface allowed the user to connect their Bose AR glasses (Fig. 5). The user was able to select the language to translate to from a provided dropdown list. It also connected to the microphone on the glasses and begin recording using scripts when the conversation started. The audio files were sent to AWS in an MPEG Audio Layer-3 (mp3) file format over the Internet for transcription and translation. Unity then received the translated mp3 audio file back from AWS. The audio file was then over the speakers in the Bose AR glasses. An example of a gesture used would be shaking your head "no" to have the translated text repeated. Double tapping the frames was used to start and stop recording.

Besides the user interface, back-end function on the web servers was necessary to support the translation application. AWS was used to host the back-end code and data for the system. AWS offered the capability to host servers and AI services that were accessed through API [8]. The back-end modules were hosted on AWS. There were three main modules used in this process. They were transcribing speech to text, translating the text to the desired language, and converting the translated text back to speech. Each of these three modules required their own API to communicate with AWS. Speech to text utilized AWS Transcribe API, and the use of an S3 bucket to save the original audio file. Text translation utilized the AWS Translation API. Text to Speech utilized the AWS Polly API. These modules were integrated with each other to provide seamless speech to speech translation. The backend code was written in PHP, since AWS has many features

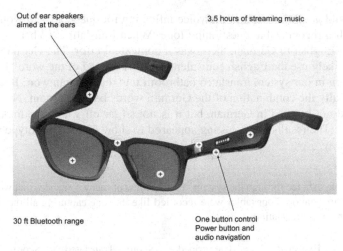

Fig. 5. (Bose AR Sunglasses Interface) Modified from www.Bose.com. This image shows the important features present in the sunglasses. The sunglasses have powerful AR features for the user to use.

and tutorials on how to manage the infrastructure in apps, with the example code being PHP [3].

3 Testing

Testing of the design was primarily focused on the accuracy of the transcription and translation, and the delay time. The method used was putting sample sentences and phrases through the system as a user would, which captured the outputs of the transcription, translation, and timestamp in a log file. Then this data was analyzed for accuracy and delay time. An important distinction to make was that a translation error would occur if there was a contextual error in the translation. People knowledgeable of the output languages were used to verify the validity of the foreign language phrases. The scoring only compared with the prior step. For example, if there was an error in transcription and the translation module correctly translated the input text, then an error would be counted for the transcription step, but no error would be counted for the translation step. The process for measuring input and output translation success was performed four times with different language combinations. The different combinations were English to German, English to Spanish, Spanish to English, and German to English. Another measurement was the time required, such as whether different language translations or varying complexity of sentences had a significant impact on translation time, or if there was little apparent impact.

4 Results

Table 1 shows examples of phrases that were tested in the system and the resulting translation. The system correctly picked up names as proper nouns. The system also correctly would get the context of the voice inflection for questions and would say the question with a recognizable questioning tone. When translating "Where is the bathroom?" from English to German, there was a contextual error. The German language would more likely use the English equivalent of toilet instead of the word "bathroom". The translation in our system translated bathroom into the German word Badezimmer, which is literally the combination of the German words bath and room. Normally this phrase would not be used in German, but it is not so far off where the understanding would be lost. The results of the testing appeared to show that the prototype worked in an effective way.

Table 1. Sample data fed through the system. This is a few of many sentences that went through each language translation. Four tables were recorded like this one capturing all the data for the system handling each translation.

Input	Expected	Transcribed Output	Score	Translated Output	Score	Time
Hello my name is…	Hallo, mein Name ist…	Hello my name is Forrest	1	Hallo, mein Name ist Forrest	1	4.052641
Where is the bathroom?	Wo ist die Toilette?/Wo ist das Klo?/Wo ist das WC?	Where is the bathroom	1	Wo ist das Badezimmer	0	3.83589
How much does that cost?	Wie viel kostet das?	How much does that cost	1	Wie viel kostet das?	1	3.860418
Where is the closest transportation?	Wo gibt es Verkehrsmittel in der Nähe?	Where is the closest transportation	1	Wo ist der nächste Transport	0	4.122393
What do you do for work?	Was machen Sie zum Beruf?	What do you do for work	1	Was tun Sie für die Arbeit	1	3.86144
Thank you, goodbye	Vielen Dank, tschüss	Thank you goodbye	1	Danke auf Wiedersehen	1	3.885468
I'm sorry (apology)	Es tut mir leid	I am sorry	1	Es tut mir leid	1	3.663461
The car is red	Das Auto ist rot	The car is red	1	Das Auto ist rot	1	3.991076

The results for speed and accuracy showed consistency in the system no matter which language combinations or phrases were used. The system's transcription success rate was 100% for English, 93.3% for Spanish, and 73.3% for German (Fig. 6). Part of the speaking and phrasing errors for Spanish and German transcription may have resulted due to the fact that the subjects speaking the phrases were not native speakers of those languages and the errors might have been mitigated if native speakers had been speaking into the system. The translation success rate for the system was also satisfactory at 80%–86% for the four language combinations.

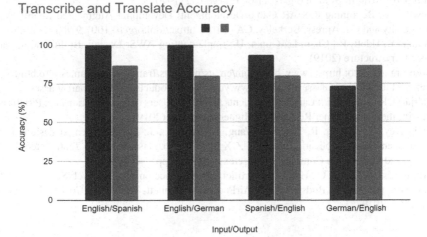

Fig. 6. Graphical representation of the accuracy of the system transcribing and translating each language.

The time for total translation was also consistent which makes the prototype a viable solution for real time communication. Table 2 show times required for translations during the testing process. The four language combinations had average translation times for the sample data that were similar. The graph also displays the consistency for the one sample of English to German translation

Table 2. Table with average translation time.

Input/Output	Avg. time (s)	Standard deviation
English/Spanish	3.90	0.12
English/German	3.90	0.13
Spanish/English	3.5	0.70
German/English	3.92	0.60

5   Conclusion and Future Direction

The translation system of passing audio files through an Amazon server to a Unity app showed promise. More development and testing is required toward having a system ready for wide application. The development of such a product would improve language translation for hands-free auditory language translation. The prototype showed reasonable response speeds and response accuracy, encouraging further development.

References

1. Min, C., Mathur, A., and Kawsar, F.: Exploring audio and kinetic sensing on earable devices, in Jun 10, 2018. https://doi.org/10.1145/3211960.3211970
2. Fowler, A.: Beginning iOS AR Game Development: Developing Augmented Reality Apps with Unity and C#. Apress, Berkeley, CA (2019). https://doi.org/10.1007/978-1-4842-3618-5
3. Wadia, Y., Udell, R., Chan, L., Gupta, U.: Implementing AWS: Design, Build, and Manage your Infrastructure (2019)
4. Bose Frames Alto: https://www.bose.com/en_us/products/frames/bose-frames-alto.html
5. Vuzix M4000 Smart Glasses: https://www.vuzix.com/products/m4000-smart-glasses
6. Valijärvi, R., Tarsoly, E.: Language Students as Critical Users of Google Translate: Pitfalls and Possibilities, Practitioner Research in Higher Education (2019)
7. McKelvey, C., Dreyer, R, Zhu, D., Wang, W., Quarles, J.: Energy-oriented designs of an augmented-reality application on a VUZIX blade smart glass. In: 2019 Tenth International Green and Sustainable Computing Conference (IGSC), pp. 1–8 (2019)
8. Tripuraneni, S., Song, C.: Hands-On Artificial Intelligence on Amazon Web Services (2019)
9. Google launches Pixel Buds, its Apple AirPods rival, Bennett, Coleman & Co. Ltd, New Delhi. Oct. 16, (2019)

Haptic Vibrations for Hearing Impaired to Experience Aspects of Live Music

Nicolas DeGuglielmo, Cesar Lobo, Edward J. Moriarty, Gloria Ma, and Douglas E. Dow(✉)

Electromechanical Engineering, School of Engineering, Wentworth Institute of Technology, Boston, MA, USA
{deguglielmon,loboc2,mag,dowd}@wit.edu

Abstract. Listening to music contributes to community building and benefits mental health. The ability of hearing-impaired and deaf individuals to experience and benefit from music is hindered. Musical instruments vibrate to generate sound. The tactile feelings of these vibrations were utilized by deaf historical figures of Beethoven and Hellen Keller. Snakes have a keen sense of tactile sensations induced by both ground-borne and airborne vibrations, and the induced neural processing occurs in both the somatic system and the auditory cortex. Congenitally deaf individuals have been shown to have an enhanced ability to process tactile information, including processing in the auditory cortex. Wearable garments and furniture have recently been developed to convert audible sounds and music into vibrations to enable hearing-impaired and deaf individuals to have some level of experience. These devices show promise, but more work needs to be done to improve the conversion of sound into haptic vibrations in a more meaningful way. The objective of our project was to develop a wearable electronic system that would extract volume and frequency features from the recent moments of live music, and to use that information to generate vibrations at multiple locations on the skin toward enabling an experience of the music. A prototype was designed and developed. Several submodule tests were conducted to evaluate functionality. A human-subject pilot test was conducted to evaluate whether the vibration pattern would relate to the music, and possibly help to distinguish types of songs that had different genre. In the test, subjects were tasked with selecting which song was being used by the systems to generate the haptic vibrations. The subjects appeared to be only slightly more accurate in their song selection than would be expected by chance. The system shows promise, but more development and testing would be required toward wider application.

Keywords: Tactile · Impact vibrations · Deafness · Wearable electronic

1 Introduction

Hearing impaired individuals have difficulty experiencing the benefits of music. Music has been used through history as a means of entertainment, community building and

T. Nakano (Ed.): BICT 2021, LNICST 403, pp. 71–86, 2021.
https://doi.org/10.1007/978-3-030-92163-7_7

expression. Music has numerous health benefits [1, 2]. The art of music has been shown to have beneficial value for humans in many ways, such as cognitive and academic improvement for children and seniors, pain relief, relaxation, and exercise [2]. However, individuals who are hearing impaired or deaf are unable to fully perceive music or reap the benefits that come with listening to music. Individuals with deafness typically would not have audio induced neural stimulation from the audio sensory receptors in the cochlea to be processed in the auditory cortex. Vibration of tactile sensors in the skin induces some neural stimulation with similarity to auditory signals [3, 4].

Before studies comparing neural activity from audio and tactile vibrations were conducted, some intuition on the connection was observed in the popular culture. Percussion and string musical instruments undergo observable vibrations as sounds are generated. The story of historical composer Ludwig Van Beethoven is well known, in which he become deaf prior to composing what arguably became his greatest symphony. During that period, he would arrange to sense tactile vibrations from a piano to perceive some aspects of his music. He would lay the piano chassis directly on the floor, sit next to it and hit the keys hard. He would also bite a rod attached to his piano to feel the vibrations through his teeth [5]. A century later, the deaf and blind historical figure of Hellen Keller would learn to listen and then to speak with an audible voice by placing her hand on her teacher's mouth and feeling the vibrations from their voice [6]. Electronic speakers can generate vibrations that can be felt by tactile sensors in the hand. The 1995 movie of *Mr. Holland's Opus* portrayed the story of a high school music teacher who had a deaf son. To allow his son to experience some aspects of music he would have his son touch or sit on an electronic speaker that would be playing the music with high volume [7].

Understanding of the physiology connecting auditory sensory systems and tactile sensory systems was aided by experiments with snakes. Snakes have both an auditory sensory system (inner ear, VIII cranial nerve) and a somatic sensory system (skin mechanoreceptors, neural signals passing through spinal cord). Both the auditory and somatic sensory systems were found to respond to airborne sound, either focused near the head or along the body [8, 9]. The snakes were found to respond both with neural activity and behaviorally in similar ways as stimulated by either airborne or ground-borne vibrations [8, 10].

Such stimulation of both auditory and tactile sensory system was not only observed in snakes, but also in humans. Brain imaging studies found neural activity in the auditory cortex induced by tactile vibration [11, 12]. This phenomenon of tactile stimulation inducing activity in the auditory cortex was found to be enhanced in congenitally deaf individuals compared with normal-hearing individuals [12, 13]. Neural plasticity following long-term auditory sensory deprivation appeared to improve the ability of the auditory cortex to process tactile information [13]. Extending from sound to music, the full experience of "hearing" music involves many neural systems, not just the cochlea and auditory cortex, but other regions as well [13, 14].

Inspired by these observations in nature, physiology and experience, some prototype physical devices have been developed toward helping hearing-impaired or deaf individuals experiences some aspects of sound and music. A Sound-Shirt was developed and marketed by CuteCircuit (https://cutecircuit.com) [15]. In this Sound-Shirt, vibration actuators woven into the fabric of the garment translated musical sounds into touch-like sensations on the wearer's back, sides, shoulders, and arms [15, 16]. The sensation of music may not be complete, but the idea shows promise to allow some aspects of the music to be experienced [16, 17]. Another prototype project placed the vibration actuators in chairs and furniture, such that when a deaf person sits within the chair, vibrations were generated to different parts of the body. The vibrations were based on certain correlations with the music, and were intended to allow the individual to experience some aspects of the music [18].

These devices that attempt to translate the audio aspects of music to haptic sensations to allow deaf individuals to experience music have limited function and success, yet show promise for the concept. More development needs to be done toward exploring the mapping of music to haptic vibrations. The purpose of our project was to explore methods to map audio waveforms to haptic vibrations within a wearable band.

2 Materials and Methods

2.1 Design

Hardware Design

A prototype was designed for the following functions. A microphone was used for audio input. Algorithms running on a microcontroller processed the audio signal to extract key features within recent small time-windows, specifically frequency and volume. These features were mapped to drive two pairs of vibration actuators within a band to be worn, allowing for haptic tactile sensations of the vibrations. The intent was for some aspects of the music to be conveyed by the pattern of vibrations on the skin.

The following components were selected for the prototype. The microphone was an Adafruit Max 9814 component (New York City, NY, USA). This microphone module had a small size and included a built-in op-amp of adjustable gain and ability to adjust the sensitivity in response to sudden changes in volume. The microcontroller was an Arduino (www.arduino.cc) Uno. As the prototype becomes further developed toward a wearable, mobile system, the Arduino Uno could be replaced with an Arduino Lilypad, more suitable for wearable electronics. The Arduino digitized the analog waveforms from the microphone, ran algorithms to process the audio signal, and output pulse width modulation (PWM) signals to motor drivers. The motor drivers were Adafruit DRV8871 modules, which generated the required 5 V and 80 mA to drive the motors. The motor for vibration was a linear resonant actuator (LRA) motor (#G0825001D, Jinlong Machinery & Electronics, Wanchai, Hong Kong). These coin motors generated vibrations that would induce noticeable tactile sensation on the skin and had a size suitable for incorporating into a wearable band.

LRAs are one of several methods to develop haptic vibrations, typically utilized for smartphones and game controllers. LRAs use magnetic fields and electrical currents to create a force, similar to the mechanisms in classic magnetic coil audio speakers. Changes in the current through the coil changes the magnetic force, causing movement of a magnetic mass. Driving the magnetic mass up and down causes the displacement of the LRA and thereby the haptic vibration.

Algorithm

A mapping is necessary between the wide range of frequencies of audio waveforms and the more limited frequencies of vibrations to be generated on the skin for haptic sensation. The range of audio frequencies detected in the cochlear and the range of haptic tactile vibration frequencies detected by mechanoreceptors in the skin are different. Audio frequencies are generally considered to have a wide range from 20 Hz to 20 kHz, but skin tactile vibration frequencies have a narrow range, generally considered to be less than 500 Hz. For example, the low frequencies generated by a bass speaker may be felt when one touches the speaker, but not when only the higher audio frequencies are being generated.

Another limitation for the mapping of frequencies is the use of the LRAs to generate the haptic sensation. Unlike a classic magnetic audio speaker that can vibrate the speaker diaphragm at a wide range of frequencies, a LRA can only vibrate near its resonant frequency. The frequency in sound waves is the pitch of the sound, thus representation of music requires generation of multiple frequency vibrations.

To overcome these limitations that tactile sensation can only sense a small portion of the lower range of audio signals, and the LRA can only vibrate at one frequency, a concept of mapping the audio information to multiple LRAs located at different locations along the skin was employed. Vibrations at one location would reflect certain of the features of the audio signal, and vibrations at another location would reflect different features of the audio. With time and learning, possibly the brain could interpret these spatially distinct vibrations as different aspects of the audio signal. Figure 1 and Fig. 2 shows such the system design for this concept.

The algorithm to process the digitized audio waveform and extract the key features ran on the Arduino. Samples of the audio signal were analyzed sequentially to extract two features: a frequency and a volume for the recent samples. The algorithm was customized and adapted from an Instructible Sample Code [19]. The customized algorithm utilized in the prototype determined a frequency value as follows. The slope between two consecutive samples was calculated. A slope of ~0 indicated a local minimum or maximum peak in the signal. The time between two consecutive peaks was used to calculate a frequency value. In parallel with the slope and frequency algorithm, an estimate of average amplitude for the most recent samples was estimated. Changes in the amplitude or volume correlated with certain aspects of the music, such as reflecting beat or rhythm.

These features of amplitude and frequency were used to determine the PWM signals to drive each LRA. The prototype had two pairs of LRA, intended to be symmetrical for the left and right side of the band. A pair consisted of two LRAs. The first LRA was driven mainly by the amplitude feature if the frequency value was below a threshold, such as 1 kHz. The second LRA was also driven by the amplitude, but only if the frequency value was above the threshold value. In this way, the spatially separate locations of vibrations on the skin would convey low frequency amplitude in one location, and high frequency amplitude in the other location. Having vibration patterns differ based on frequency could deliver a more dynamic experience to the user and potentially improve the ability to distinguish between types of music or songs.

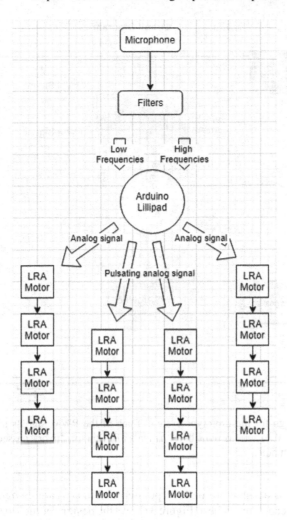

Fig. 1. Concept diagram of multiple vibrations at spatially different locations on the skin to map different features of the audio signal. Audio signals from a microphone, pass through filters to control the frequency content. Then the signal is digitized and processed by algorithms in the Arduino microcontroller. These extracted features are used to drive the LRA to generate vibrations at spatially different locations on the skin for haptic tactile sensations.

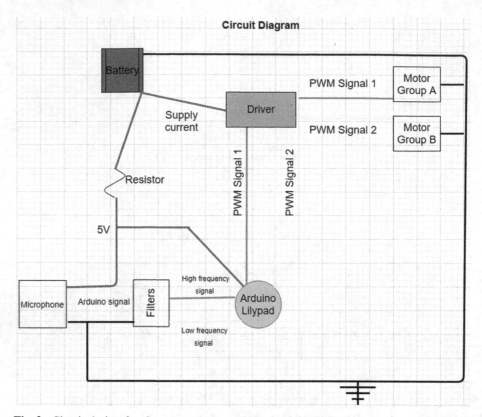

Fig. 2. Circuit design for the prototype system. Each of the PWM signals transmitted by the Arduino was mapped to one of the filtered signals. The motor drivers passed the appropriate signals to drive the motors.

The system was designed to be incorporated within a wearable band that could be worn around the head, arm or chest. Figure 3 shows the design for the wearable band. The base material was a flexible fabric with adjustable Velcro straps. The band was designed to incorporate the electronics: microphone, Arduino, battery, motor drivers and two pairs of two LRA motors each.

For the prototype that was utilized for the testing reported in this paper, Fig. 4 shows the circuit connections of the primary components and the wearable band. The band was composed of a 2-inch elastic fabric strip with pockets sewn into either side. The electronic components were sewn into these pockets in the arrangement seen in Fig. 4. The two pairs of LRA motors were wired to their respective driver modules. Wires connected the driver channel input pins and common ground wire to an external Arduino Uno and microphone assembly. The prototype band was made to be positioned on the head at the forehead and temple position. The adjustable Velcro strap allowed for proper fitting.

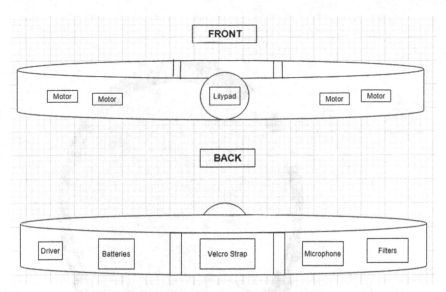

Fig. 3. Concept diagram of the headband design. All components were sewn into elastic fabric pockets with the necessary wires running around the band. Motors were placed symmetrically around the wearer's temple with other components placed with the goal of optimizing wire length

3   Testing and Results

3.1   Frequency Detection Test

The ability of the algorithm to extract frequency values was tested by having the input sound be a constant tone of known frequency. In one trial, a tone of 440 Hz frequency was played on a sound wave generation smartphone app with the speaker of the phone placed near the microphone. In a second trial, a tone of 950 Hz was used. The results are shown in Fig. 5 with the 450 Hz trial on the left, and the 950 Hz trial on the right. The upper plots show the input waveform of the tone as displayed by the Arduino serial plotting tool. The plots on the bottom show the frequency values determined by the algorithm. In both trials, the correct frequency was often found, as shown by the upper values of ~450 on the bottom left plot and ~950 on the bottom right plot. Both also had many erroneous values of lower frequencies. This was considered acceptable for the prototype in that the algorithm to characterized recent frequency values was intentionally simple in order to not bog down the processor and allow for almost real-time characterization. The correct value, or lower values were found throughout these trials.

3.2   Volume Detection Test

While the algorithm was determining values for frequency, amplitude was also estimated to find the volume level for the recent samples. Within the code, the absolute value of the amplitude of the recent samples was calculated. This amplitude value indicated a volume level. To test this detection of volume, periodic short bursts of noise was generated near

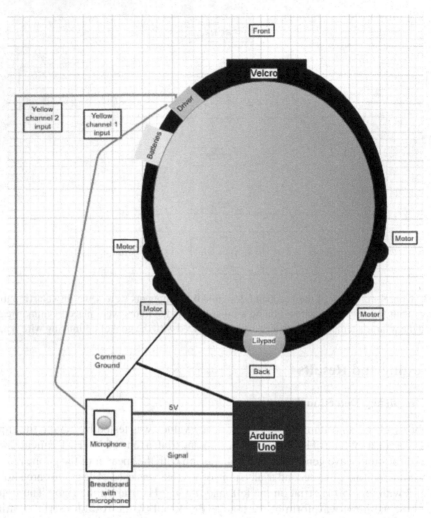

Fig. 4. Top view of the prototype. The microphone was left separate from the main headband and powered by the Arduino Uno for convenience in testing procedures and to avoid power distribution issues.

the microphone with periods of quiet between these bursts. The resulting plot of volume is shown in Fig. 5. The algorithm was able to determine an estimate of the volume of recent audio samples.

3.3 LRA Vibration for Volume Test

The volume value from the algorithm for the audio signal was used to set the duty cycle for the PWM signal to drive the motor. A higher volume value was mapped to a higher duty cycle. The higher duty cycle drove the LRA to vibrate for a larger percent of the time. The tactile sensation of vibrations with a higher duty cycle could be interpreted as

Table 1. Frequency detection for tones of constant frequency. The left half shows results for a trial of a tone having 450 Hz. The right half shows results for a trial of a tone having 950 Hz. The upper plots are the input waveforms as plotted by an Arduino plotting function. The bottom half shows the frequency. In both cases, the correct frequency was often detected as shown by the high values (~450 on the bottom left and ~950 on the bottom right), with other erroneous values having lower frequencies.

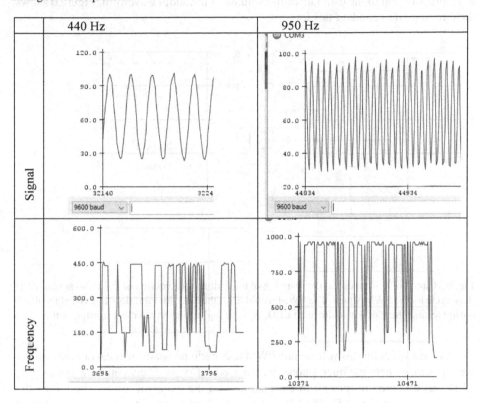

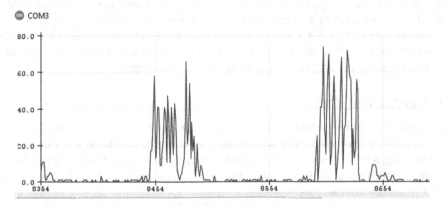

Fig. 5. Volume level of the recent samples as estimated by the algorithm.

a higher volume, and vibrations with a lower duty cycle could be interpreted as a lower volume. To have visual feedback during testing, the PWM signal was also feed to an LED. A higher duty cycle would result in a visually brighter LED, and a lower duty cycle would result in a dimmer LED. The PWM was the same one intended to drive the LRA for vibration. The mapping program on the Arduino set the duty cycle of the PWM to be proportional to the estimate of the volume of the audio waveform. Figure 6 shows the circuit used to conduct this test.

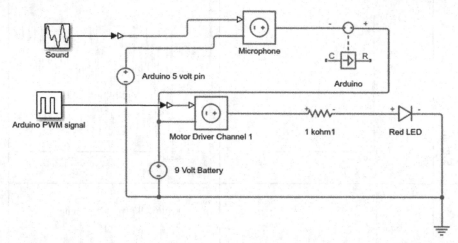

Fig. 6. Circuit of system to test the mapping of the estimate for volume of the audio signal to the duty cycle for the PWM signal. The PWM would drive the LRA for vibration. For visual feedback during testing, the PWM also drove an LED, thus the higher the duty cycle, the brighter the LED.

As audio input for this volume and PWM test, periodic bursts of vocal popping noises were generated near the microphone, with periods of silence between these bursts. The LED was observed to turn on in correlation with the bursts of noise. The louder the burst, the brighter the LED. The LRA was gently held by the fingers during this test as subjective feedback of haptic feedback. Figure 7 shows a photograph of the testing setup with LED and the LRA being held by the fingers for haptic detection of vibration.

The intensity of the LRA vibration was found to be correlated with the timing of the bursts of noise. The louder the bursts, the stronger the feeling of vibration on the fingers. This functionality was considered acceptable for the prototype.

3.4   Low Pass Filter Test

The design of the prototype had two LRAs in each pair. One LRA was to vibrate according to the volume of lower frequency sounds and the other was to vibrate according to the volume of higher frequency sounds within the music. A low pass filter (LPF) and high pass filter (HPF) were used to separate the musical content into lower frequency and higher frequency components.

A passive, single order, resistor-capacitor (RC) filter was constructed with a cutoff frequency of 339 Hz. Figure 8 shows a diagram of the LPF circuit used for testing. The

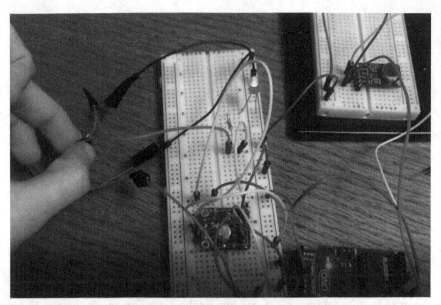

Fig. 7. Photograph of prototype during testing of the correlation between volume of the sound and duty cycle of the PWM signal. The large the duty cycle, the brighter the LED would be illuminated and the more intense the feeling of haptic vibration as felt by the fingers gently holding the LRA.

goal of the test was to determine the level of attenuation of the signal by the filter, and the estimation by the algorithm of frequency and volume. This was done for two tones. One tone at 300 Hz, which was below the cut-off frequency. The other tone was at 2000 Hz, which was above the cut off frequency. The voltage amplitude of the filter output could be used in future versions of the algorithm as a boundary at which the program would ignore the signal and not determine the frequency values.

For the tested LPF with a cutoff frequency of 339 Hz, the LPF was observed to atten uate by ~33% the average amplitude of the 2000 Hz tone compared to the amplitude of the 300 Hz tone. According to the theoretical Bode plot for such a LPF, even higher frequency tones would be expected to be attenuated even more. In either case the algo- rithm was still able to detect the frequency value. Such a LPF and a HPF could be used to separate the frequency content of music into lower and higher frequency components, but the separation would be sharp, and much overlap in frequency content would occur. In the further testing of the prototype for this paper, the filter was not utilized.

3.5 Dual PWM for Two LRAs Test

This was a test to see if the system could generate two different PWM signals to drive the two LRAs. The algorithm on the Arduino calculated an estimate of the frequency and volume of recent samples of the audio waveform. Then the value for volume (regardless of the corresponding frequency) was used to generate the PWM to drive the first LRA. Then, only the volume that corresponded to frequency values above the threshold of 1 kHz was used to generate the PWM to drive the second LRA. In intension was for

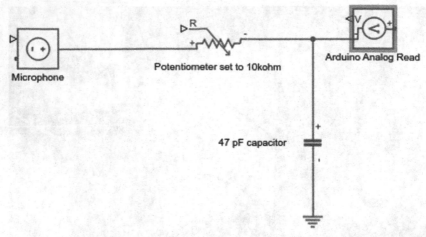

Fig. 8. Passive LPF circuit composed a resistor and a capacitor. The cutoff frequency of the filter was 339 Hz.

the motors to be driven differently depending on the corresponding frequency. Figure 9 shows a diagram of the system used for this test.

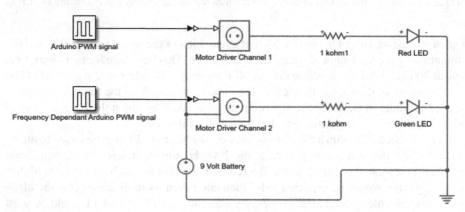

Fig. 9. Dual channel test circuit. Cannel 1 receives a PWM signal mapped to the amplitude of the signal transmitted from the microphone. Channel 2 receives the same PWM signal only when the program indicates that the sound is of greater frequency than 1 kHz.

The LED's responded as intended with the red LED varying in brightness with the sound's volume and the green LED only activating when its respective frequency is played. The green LED pulsated due to the clipping of the calculated frequency. This clipping, as displayed in Tables 1 and 2, is caused by occasional inaccuracies in the algorithm's calculation. The incorrect frequency values fell below the green LED's minimum frequency thus prompting the green LED to turn off for that instant. The periodic deactivation of the green LED mimicked the pulsing behavior that was intended for the secondary LRA, thus this clipping was utilized to pulsate the high frequency motors.

3.6 Response to Music

The purpose of this test was to observe the system's response to music. Three songs of different genres were played in a large room on television speakers and the program's interpretation of the sound's volume and frequency. These specific songs were selected because they varied from one another in tempo and rhythm which would likely be represented in different volume and frequency patterns calculated by the algorithm. The test examined how the volume of the input music affected the boundaries of the recorded volume and frequency. Each song was played once at higher volume and a second time at a lower volume with a sample of the transmitted signal, amplitude and frequency collected at a random point in the song. The results were intended to establish reasonable boundaries of volume and frequency that would be calculated by the algorithm when responding to live music.

The songs selected for the test were as follows:
  Song 1: "Through the fire and flames" by Dragonforce
  Genre: Hard Rock

  Song 2: Ben Matthews's cover of "Don't you leave me here" by Jelly Roll Morton
  Genre: Smooth Jazz

  Song 3: "Came a long way" by Dark Chocolates
  Genre: Rap

Table 2. Amplitude and frequency of music samples

Song	Peak amplitude		Frequency (Hz)	
	High speaker volume	Low speaker volume	High speaker volume	Low speaker volume
1	35	30	100–6400	200–4200
2	40	35	65–1900	190–3800
3	40	32	300–2000	300–4700

The algorithm running on the Arduino Uno could not detect a volume above 40 or a frequency above 6500 Hz in any of the song samples. These values will be used as the maximum boundaries of these parameters in the program.

Table 3. Prototype test results

Subject 1

		1	2	3	4	5	Accuracy
Control	Played	2	3	1	3	2	0/5
	Reported	Unheard	1	Unheard	Unheard	Unheard	
With headband	Played	2	1	3	2	2	3/5
	Reported	2	1	3	3	3	

Subject 2

		1	2	3	4	5	Accuracy
Control	Played	3	2	1	3	2	1/5
	Reported	Unheard	Unheard	Unheard	Unheard	2	
With headband	Played	3	1	3	2	3	2/5
	Reported	3	1	1	3	2	

Subject 3

		1	2	3	4	5	Accuracy
Control	Played	1	3	2	3	1	0/5
	Reported	Unheard	Unheard	Unheard	Unheard	Unheard	
With headband	Played	3	2	1	3	1	0/5
	Reported	Unheard	Unheard	Unheard	Unheard	Unheard	

3.7 Song Differentiation Test

The completed prototype utilized two pairs of parallel motors. Each pair was connected to an independent channel of the motor driver. The driver's voltage output for each channel was controlled by the Arduino which transmitted the amplitude mapped PWM signal to each channel's signal input pin. One channel only transferred the PWM signal when the sound's frequency was above 1 kHz. The driver drew current from the lithium polyester battery to drive the motors in accordance with the signals transmitted by the Arduino. The microphone remained on a separate board for testing purposes but was attached to the appropriate Arduino pins with long wires as shown in Fig. 4.

The three test subjects were not deaf or hearing-impaired, and were able to hear sounds in their daily life. However, during the haptic tests they could only feel the vibrations generated by the LRAs in a headband they wore. They wore noise canceling headphones to prevent being able to directly hear the song being played. Before testing procedures began, the test subjects were introduced to the three songs to familiarize them with their respective tempos, rhythms and tones. In this test, each subject wore the headband and a pair of active noise canceling headphones to simulate impaired hearing, so that they would not directly hear the song. In the control test, the music was played from a smartphone and the subject was asked to guess which of the 3 selected songs was currently playing. If the subject could discern the music without the influence of the headband, the music volume was reduced. When a volume at which the subject could not discern the song at least 4/5 times was identified, the vibration program's maximum volume was adjusted to that volume. The subject would then be asked to guess the song being played at that same volume with the assistance of the active headband. By chance alone, a subject under these conditions was expected to guess the correct song 1.6/5 times. After the tests were conducted, the subjects were asked for suggestions on how to improve the comfort and appearance of the prototype (Table 3).

With the assistance of the headband, test subject 1 was able to guess the correct song 3/5 times while test subject 2 was able to guess the correct song 2/5 times. Songs 2 and 3 were most often confused for each other and as noted by the test subjects, felt similar or too similar to each other to consistently distinguish through vibrations. The similarity in vibration patterns between these two songs could be due to the similarity of the input parameters noted Table 2. When asked for suggestions on improving the comfort and appearance of the band, subjects 1 and 2 noted that the wires were obstructing their faces and were more comfortable when they put on the headset with the motors and wires against the back of their heads. Test subject 3 was unable to distinguish any of the songs correctly claiming that the motors were experiencing high levels of noise and did not vary in intensity. This could be the result of the sensitivity changes that were made to cater to subject 3's ability to hear which greatly differed from that of subjects 1 and 2.

4 Conclusion and Future Directions

The prototype system produced mixed results during the pilot test for assistance with discerning music songs. A more dynamic user experience would be required to accurately represent music, thus, the inputs and outputs of the system should be develop more. The primary testing had the vibration band worn on the head, but other locations were not yet tested, such as around the chest or arms. Future designs could incorporate improved filters to separate frequency components of the music. Motors with a wider range of vibration frequencies, such as Piezo actuators could be utilized. Reorientation of the components within the wearable band would also improve comfort and practical aspects of the system.

Further development on mapping of aspects of live music with haptic vibrations is necessary toward wider application. Such systems could enhance the ability of hearing impaired and deaf individuals to experience aspects of live music. Such experience might improve feelings of community with others listening to the music and potentially allow for other benefits of music on mental health.

References

1. Anvari, S.H., Trainor, L.J., Woodside, J., Levy, B.A.: Relations among musical skills, phonological processing, and early reading ability in preschool children. J. Exp. Child Psychol. **83**(2), 111–130 (2002)
2. Mazaheryazdi, M., Aghasoleimani, M., Karimi, M., Arjmand, P.: Perception of musical emotion in the students with cognitive and acquired hearing loss. Iran J. Child Neurol. Spring **12**, 41 (2018)

3. Neves, J.: Music to my eyes... conveying music in subtitling for the deaf and the hard of hearing. 123–146 (2010)
4. Johnson, M.S.: Composing music more accessible to the hearing impaired (2009)
5. Perciaccante, A., Coralli, A., Bauman, N.G.: Beethoven: his hearing loss and his hearing aids. 41, 1305–1308 (2020)
6. Rogow, L.: The unconquered: Helen Keller in her story. 48, 252–254 (1954)
7. Henrek, S.: Mr. Holland's Opus (1995).
8. Hartline, P.H.: Physiological basis for detection of sound and vibration in snakes. J. Exp. Biol. 54, 349–371 (1971)
9. Proske, U.: Vibration-sensitive mechanoreceptors in snake skin. Exp. Neurol. 23, 187–194 (1969)
10. Young, B.A.: Snake bioacoustics: toward a richer understanding of the behavioral ecology of snakes. Q. Rev. Biol. 78, 303–325 (2003)
11. Schürmann, M., Caetano, G., Hlushchuk, Y., Jousmäki, V., Hari, R.: Touch activates human auditory cortex. Neuroimage 30, 1325–1331 (2006)
12. Auer, Jr. E.T., Bernstein, L.E., Sungkarat, W., Singh, M.: Vibrotactile activation of the auditory cortices in deaf versus hearing adults. 18, 645 (2007)
13. Levänen, S., Hamdorf, D.: Feeling vibrations: enhanced tactile sensitivity in congenitally deaf humans. Neurosci. Lett. 301, 75–77 (2001)
14. Koelsch, S.: Neural basics of music perception, melody, harmony, and timbre. In: Brain & Music. Wiley (2012)
15. Hawthorn, A.: This shirt enables the deaf to feel music. Psychology Today. https://www.psychologytoday.com/us/blog/the-sensory-revolution/202001/shirt-enables-the-deaf-feel-music (2020)
16. Motola-Barnes, M.: Haptic actuators: comparing piezo to ERM and LRA. https://blog.piezo.com/haptic-actuators-comparing-piezo-erm-lra. Accessed April 2021
17. Grossi, A.D.: The sound shirt (2020)
18. Jack, R., McPherson, A., Stockman, T.: Designing tactile musical devices with and for deaf users: a case study. (2015)
19. Amandaghassaei: Arduino frequency detection. Instructables: Circuits (2017)

Experiments on *Pause and Go* State Estimation and Control with Uncertain Sensors Feedback

Violet Mwaffo[1]([✉]) [iD], Jackson S. Curry[2], Francesco Lo Iudice[3] [iD], and Pietro DeLellis[3] [iD]

[1] United States Naval Academy, Annapolis, MD, USA
mwaffo@usna.edu
[2] University of Colorado Boulder, Boulder, CO, USA
jackson.curry@colorado.edu
[3] University of Naples Federico II, Naples, Italy
{francesco.loiudice,pietro.delellis}@unina.it

Abstract. A bio-inspired state estimation and control algorithm is experimentally tested to autonomously balance a team of robots on a circle. In this control scheme inspired from the social behavior of some insects species, a leader is elected randomly and periodically moves at a constant angular speed. The followers triggered by the leader motion, implement a decentralized and non-cooperative state estimation and control algorithm using uncertain and noisy proximity sensor measurements. Individuals in the team are immobile during the *pause* sequence to gather and process proximity distances, identify closer neighbors, and estimate their relative phase distances. During the *go* sequence, they either accelerate to achieve the desired spacing from closer neighbors, or move at a constant angular speed in phase with the leader. The scheme is tested on caster wheeled robots equipped with a rotating sonar platform to get forward and backward distances and is shown capable to balance the team of robots even in the presence of false readings or intermittent measurements. Further, at steady-state, the team of robots is capable to self balance in the absence of sensor feedback.

Keywords: Bio-inspired · State estimation and control · Autonomous systems

1 Introduction

Coordinating multiple vehicles is becoming pervasive for potential applications in collective search and rescue missions [3], disaster relief and management

Supported by the United States Naval Academy and by the program "STAR 2018" of the University of Naples Federico II and Compagnia di San Paolo, Istituto Banco di Napoli - Fondazione, project ACROSS.

T. Nakano (Ed.): BICT 2021, LNICST 403, pp. 87–101, 2021.
https://doi.org/10.1007/978-3-030-92163-7_8

systems [18], and surveillance and reconnaissance missions [1]. To achieve the autonomous coordination of multiple vehicles, formation control algorithms have explored path or way points tracking [28], patterns configuration [25], or moving through constrained environment [8]. Among the existing algorithms, leader-follower formation control schemes [7] are popular and cost-efficient. Leader-follower relationships are commonly hypothesized to be fundamental in explaining the emergence of collective behaviour in several natural organisms. Indeed, leaders are often considered as informed individuals guiding other group members toward set locations. In insect groups for example, leadership might benefit the group during foraging or migration [10]. As such, leaders-followers relationships are central to the understanding of the interaction between group members. For robotic applications, only a handful of agents, denoted leaders, need to utilize navigation tools such as Simultaneous Localization and Mapping (SLAM) [23] or Differential Global Positioning System (DGPS) [11] to drive the followers towards the desired path, whereas the followers only rely on cheap proximity sensors such as sonar, infrared, or camera to stay aligned with the rest of the team. These sensors have very short range and their performance might significantly deteriorate in changing or cluttered environments [26].

For real world applications, decentralized and local communication protocols are often preferred to centralized approaches as they typically require less computation resources and sensing capabilities [16]. In case of limited communication due to cluttered environments or the medium restraining wireless communications, each robot has to rely on its own sensing capabilities [21]. To improve the accuracy of both relative and absolute localization methods, research efforts have proposed robust estimation methods [5], non-linear Kalman filter [19], or the use of diversified pool of sensors [22]. In localization problems [20], individual robots have to independently regulate their dynamics using either relative positions with respect to internal kinematics [9], or absolute position with respect to a reference frame [19]. In these problems, few works have addressed the case of the absent or intermittent feedback data that might arise when proximity sensors are employed, in the presence of a cluttered environment, in underwater applications, or in case of varying light or air conditions [4].

Here, we depart from a bio-inspired estimation and control scheme [24] and illustrate how a recent state estimation and control algorithm [12,13] can be implemented on ground robots. Specifically, here we focus on the cheapest implementation in which the robots are equipped with ultrasonic sensors, whose limited accuracy need to be compensated by a synergistic design of the estimator and controller. The team of robots includes a single leader and several followers implementing the state estimation and control algorithm based on noisy and intermittent proximity distances to predict their relative position with their closest pursuant on the circle. This decentralized non-cooperative approach is implemented in a *pause-and-go* scheme, where, during the *pause*, followers recursively estimate their relative angular position to the robot ahead or behind on the circle and during the *go*, followers either accelerate or move at a constant speed to appropriately space from nearby robots. The *pause-and-go* behavior has been observed in some insect species [2] that adopt this strategy during foraging or social interaction, or to better appreciate the closeness or the alignment to a distant target or to a conspecific [27].

Outline of the Paper: Section 2 describes the formation control problem and the state estimation and control algorithm. Section 3 depicts the experimental setup and the trials. Section 4 present the results and Sect. 5 concludes the work.

2 State Estimation and Control Algorithm

2.1 Problem Statement

We consider a group of $i = 1, ..., N$ mobile robots moving on a circle of radius $R > 0$ and updating their angular position $\theta_i(k)$ at each discrete time instant k as:

$$\theta_i(k+1) = \begin{cases} \theta_i(k) + u_i(k), & \text{if } (k/p) \in \mathbb{N}, \\ \theta_i(k), & \text{otherwise}, \end{cases} \tag{1}$$

where $u_i(k)$ is the control input implemented in a *pause-and-go* fashion and $p \in \mathbb{N}$ is an integer corresponding to the number of time steps needed to perform the measurements and estimations.

The control goal is to find $u_i(k)$ in order to set the pace of the group to the a desired angular speed ω_{ref} every p time steps, and coordinate the team of robots in a balanced formation on the circle. Formally, this translates into achieving an ε bounded formation [21], that is,

$$\limsup_{k \to +\infty} |\xi_{ij}(k) - 2\pi/N| \leq \varepsilon, \tag{2}$$

for given any pair of consecutive agents (i, j), where $\xi_{ij}(k) := \text{rem}(\theta_i(k) - \theta_j(k))$ is the relative phase between robots i and j[1]; the quantity $2\pi/N$ is the desired angular spacing between consecutive agents, and ε is the formation error.

For the team of robots defined above, we consider the challenging scenario where (i) each robot is equipped with a proximity sensor with a limited range r_v that corresponds to a *visibility cone* $\varphi_v = 2 \arcsin(\rho_v/2R)$ along the circle; (ii) the on-board computer power can only allow to collect intermittent measurements every p steps, where p is the time required to acquire and to process the information; (iii) each robot, say i, obtains a noisy measurement $\tilde{d}_{ij}$ (affected by a bounded noise δ_{ij}) of the distance d_{ij} from a robot $j \neq i$ only if $d_{ij} \leq r_v$; (iv) the robots have no mean to uncover the identity and relative order (ahead or behind)[2] with respect to closer neighbors.

The above constraints imply that pairwise proximity distance satisfies:

$$\tilde{d}_{ij}(k) = \begin{cases} d_{ij}(k) + \delta_{ij}(k), & \text{if } d_{ij}(k) \leq r_v \wedge (k/p) \in \mathbb{N}, \\ \text{n.a.}, & \text{otherwise}, \end{cases} \tag{3}$$

---

[1] $\text{rem}(z)$ denotes the unique solution for r to the equation $z = 2\pi w + r$, where $-\pi \leq r < \pi$, $w \in \mathbb{Z}$.

[2] We say that robot i is ahead of j at time k if $\xi_{ij}(k) > 0$, otherwise i is behind j.

where the measurement error verifies $|\delta_{ij}| \leq \delta_{max}$ for a given positive scalar δ_{max}, and n.a. stands for no measurement available. We comment that the control objective in (2) cannot be fulfilled by traditional control schemes considering the limitations on the available measurements summarized by Eq. 3.

Introducing the relative phase distance $\varphi_{ij}(k) := |\xi_{ij}(k)|$ between robots i and j, from Eq. (3), we know that[3]

$$\varphi_{ij}(kp) \in \begin{cases} \Upsilon_{ij}(kp), & \text{if } \varphi_{ij}(kp) \leq \varphi_v, \\ \overline{\Upsilon}, & \text{otherwise,} \end{cases} \qquad (4)$$

where $\overline{\Upsilon} := (\varphi_v, \pi]$, and

$$\Upsilon_{ij}(kp) := [\max\{\tilde{\varphi}_{ij}(kp) - \varphi_{max}, 0\}, \min\{\tilde{\varphi}_{ij}(kp) + \varphi_{max}, \varphi_v\}], \qquad (5)$$

with $\tilde{\varphi}_{ij}(kp) = 2\arcsin(\tilde{d}_{ij}(kp)/2R)$ defining the phase distance corresponding to the measured Euclidean distance $\tilde{d}_{ij}(kp)$; $\varphi_{max} := 2\arcsin((r_v + \delta_{max})/2R) - 2\arcsin(r_v/2R)$ is the maximum uncertainty associated to $\tilde{\psi}_{ij}(kp)$. Note that from (4), albeit the relative position of the closer robot is unknown, we are aware that ξ_{ij} belongs to two uncertainty intervals (one in $[0, \pi)$ and the other in $[-\pi, 0)$) with width $\Gamma_1^{ij}(kp)$ and $\Gamma_2^{ij}(kp)$. Using the information coming from (4) and the knowledge of individual dynamics in (1), it is possible to shrunk the width of these uncertainty intervals. Note that the hull $H_{ij}(kp)$ of the multi-interval $\Gamma_{ij}(kp) = \Gamma_1^{ij}(kp) \cup \Gamma_2^{ij}(kp)$ is an overestimate of the uncertainty on $\xi_{ij}(kp)$. The reader can refer to [21] for basic properties and operations on intervals, which will be used when designing the control and estimation strategy presented below.

2.2 Control Strategy

Here, we leverage the synergistic control and estimation strategy first proposed in [21], which prescribes the random election of a leader, w.l.o.g. a robot, whose control law $u_1(k) = \omega_{ref}$, and thus sets the pace for the multi-agent system, while the followers $i = 2, ..., N$ implement a three-level bang-bang control law with initial value $u_i(0) = 0$. Initially, none of the robots is in the visibility range of another robot, the relative motion of the leader will determine a time-instant in which it will enter the *visibility cone* of robot $i = 2$. In that case, robot $i = 2$ starts implementing a prediction-correction algorithm to obtain an estimate $\hat{\Theta}_{21}$ of the relative phase ξ_{12}. Using this estimate, the control strategy can be activated resulting is either one of these two actions:

1. if node 1 is closer than the desired spacing $2\pi/N$, the control law u_2 is set to $\omega_{ref} + c$, with $c > 0$ introduced to move robot 2 faster than the leader, thereby incrementing the spacing distance between nodes 1 and 2;

---

[3] For simplicity, given the *pause-and-go* implementation, the measured distance and related quantities will be only defined at time instants kp, with k being an integer.

2. if the desired distance is achieved, the control input u_2 is switched to the desired speed ω_{ref}.

In turn, robot $i = 2$ while moving will eventually enters the *visibility cone* of robot 3 resulting into a sequential repetition of the above steps for all pairs of consecutive robots until the ε-balanced formation is achieved.

Note that at time kp, robot i possesses an interval estimate $H_{ij}(kp|kp)^4$ of the relative phase angle $\Theta_{ij}(kp)$ with respect to neighbors robot j. Introducing $\underline{H}$ and $\bar{H}$ as the infimum and supremum of an interval H respectively, at time kp, when the conditions

$$
\begin{aligned}
&\underline{H}_{i,i-1}(kp|kp) > 0, \\
&\bar{H}_{i,i-1}(kp|kp) < \underline{\Gamma}_1^{ij}(kp|kp), \text{for all } j \neq i-1
\end{aligned}
\tag{6}
$$

are both satisfied, robot i can unambiguously concludes that $j = i - 1$ is its closest pursuant.

The multi-level control strategy implemented by the team of robots can be summarized at any time instant k as:

$$
u_1(k) = \omega_{\text{ref}}, \qquad \text{for leader } i = 1
\tag{7a}
$$

$$
u_i(k) = \begin{cases} \omega_{\text{ref}} + c, & \text{if } \hat{\Theta}_{i,i-1}(k) < 2\pi/N \text{and } k \geq k_i, \\ \omega_{\text{ref}}, & \text{if } \hat{\Theta}_{i,i-1}(k) \geq 2\pi/N \text{and } k \geq k_i, \quad \text{for followers } i \geq 2 \\ 0, & \text{otherwise}, \end{cases}
\tag{7b}
$$

where k_i is the first time-instant that robot i is capable to identify its closest pursuant, that is, the first time instant such that (6) holds, and for any $i = 2, \ldots, N$, $\hat{\Theta}_{i,i-1}(l_0)$ is selected as $\bar{H}_{i,i-1}(\lfloor k/p \rfloor | \lfloor k/p \rfloor)$. In what follows, we introduce the estimation strategy to update the multi-interval $\Gamma_{ij}(kp|kp)$ upon which H_{ij} is computed.

2.3   Estimation Algorithm

The estimation strategy leverages the three-level bang-bang structure of the control law to perform an interval estimate $\hat{u}_j^i(k)$ of the input acting on node $i \geq 2$ at time k as:

$$
\hat{u}_j^i(k) = \begin{cases} \omega_{\text{ref}}, & \text{if } k \geq k_i, d_{ij} > r_v, j = i-1, \\ [\omega_{\text{ref}}, \omega_{\text{ref}} + c], & \text{if } k \geq k_i, d_{ij} \leq r_v, j = i-1, \\ [0, \omega_{\text{ref}} + c], & \text{otherwise}. \end{cases}
\tag{8}
$$

Each robot requires p time instants to process the measurements and compute the next control input, thereby the uncertainty on the interval estimation of ξ_{ij}, denoted Γ_{ij}, is updated every p steps using (4) for all $i = 2, \ldots, N, j \neq i$ starting from the initialization $\Gamma_{ij}(0| - p) = [-\pi, \pi)$ as:

---

^{4} The notation $|kp$ indicates that agent i has used all information collected until kp.

$$\Gamma_{ij}(kp|kp) = \begin{cases} \emptyset, & \text{if } kp \geq k_i, j \neq i-1, \\ \Gamma_{ij}(kp|(k-1)p) \cap (\overline{\Upsilon} \cup -\overline{\Upsilon}), & \text{if } kp \geq k_i, d_{ij}(kp) > \rho_v, j = i-1, \\ \overline{\Upsilon} \cup -\overline{\Upsilon}, & \text{if } kp < k_i, d_{ij}(kp) > \rho_v, \\ \Gamma_{ij}(kp|(k-1)p) \cap (\Upsilon_{ij}(kp) \cup -\Upsilon_{ij}(kp)), & \text{otherwise}, \end{cases}$$

$$\tag{9}$$

for all $k > 0$. Equation (9) prescribes that, prior to time instant k_i, when no measurement is available, robot i does not evaluate the intersection $\Gamma_{ij}(kp|(k-1)p) \cap (\Upsilon_{ij}(kp) \cup -\Upsilon_{ij}(kp))$. Thus, when $d_{ij}(kp) > r_v$, robot i just sets $\Gamma_{ij}(kp|kp) = (\Upsilon_{ij}(kp) \cup -\Upsilon_{ij}(kp))$. After time k_i, robot i stops estimating the position of all other agents except its closest pursuant $i-1$.

Algorithms 1 and 2 report a schematic implementation of the estimator (9). The uncertainty interval $\Gamma_{ij}(kp|kp)$ is recursively estimated p steps ahead as:

$$\Gamma_{ij}((k+1)p|kp) = \Gamma_{ij}(kp|kp) + \hat{u}_{ij}(kp), \text{ for all } j \neq i, \tag{10}$$

where robot i's estimate of the relative input with respect to j is $\hat{u}_{ij}(kp) := u_i(kp) - \hat{u}_j^i(kp)$ where $\hat{u}_j^i(kp)$ is given in (8).

The convergence of the *pause-and-go* estimation and control strategy is a particular case of the algorithm proposed in [21] and the convergence of the proposed scheme can be established using a similar procedure as in [21] while taking into account the periodical activation of the control scheme defined in (7a)–(10). In particular, an upper bound $\tilde{k}_i$ of the convergence time k_i can be determined.

Proposition 1 [24]. *For the multi-robot system in (1), If*

1. $|\Theta_{ij}(0)| \in [\min\{4\varphi_{\max} + 2\omega_{\text{ref}} + 2c, \varphi_v\}, \pi]$, *for all* $i = 1, \ldots, N$, $i \neq j$;
2. $2(\omega_{\text{ref}} + c) < \varphi_v$;
3. $\omega_{\text{ref}} > 0$ *and* $0 < c < \epsilon/(N-1)$,

then there exist $k_2, \ldots, k_N < +\infty$. *In addition,* $k_i \leq \tilde{k}_i$, *where*

$$\tilde{k}_i = \begin{cases} p \lceil (\theta_2(0) - \theta_1(0) - 2(\omega_{\text{ref}} + c))/\omega_{\text{ref}} \rceil, & \text{if } i = 2, \\ k_{i-1} + p \lceil (\Theta_{i,i-1}(0) - \varphi_v)/\omega_{\text{ref}} \rceil, & \text{if } i \neq 2 \wedge \Theta_{i,i-1}(0) > \varphi_v, \\ k_{i-1} + p \lceil (\theta_i(0) - \theta_{i-1}(0) - \varphi_v)/\omega_{\text{ref}} \rceil, & \text{if } i \neq 2 \wedge \Theta_{i,i-1}(0) < 0, \\ k_{i-1} + p \lceil (\Theta_{i,i-1}(k_{i-1}) - 4\varphi_{\max})/\omega_{\text{ref}} \rceil, & \text{otherwise}. \end{cases}$$

$$\tag{11}$$

The results in [21] has also been adapted to provide sufficient conditions to achieve a ε-balanced formation in the *pause-and-go* estimation and control strategy can be obtained as

Proposition 2 [24]. *For the multi-robot system* (1), *If there exist* $\varepsilon > 0$ *such that:*

1. $|\Theta_{ij}(0)| \in [\min\{4\varphi_{\max} + 2\omega_{\text{ref}} + 2c, \varphi_v\}, \pi]$, *for all* $i = 1, \ldots, N$, $i \neq j$;
2. $2(\omega_{\text{ref}} + c) < \varphi_v$;
3. $\omega_{\text{ref}} > 0$ *and* $0 < c < \varepsilon/(N-1)$;

then (i) the estimation and control strategy in (7a)–(10) *achieve a ε-balanced formation, that is* $\lim_{k \to +\infty} |\Theta_{i,i+1}(k) - 2\pi/N| \leq c$, *and in addition, (ii) the relative phase* $\Theta_{i,i+1}(k)$ *converges in finite time* $k_i^c \leq \bar{k}_i^c$ *for a given* $k_i^c \in \mathbb{N}$, *for all pair of robots* $(i, i+1)$ *with* $i = 1, \ldots, N - 1$.

Remark 1. From the above propositions, the control parameter c can be carefully selected to regulate the trade-off between accuracy and convergence speed. In particular, smaller values of c might improve the accuracy of the control scheme while increasing the convergence time. This is in particular true for larger group sizes where individuals might be required to move at a slow pace to avoid collision or motion jamming. However, for larger team sizes, it is more likely that individual robots can perceive each other, hindering a key feature of our control scheme which assumes that, at steady state, the range of visibility is lower than the desired spacing distance. Note that, in that case, alternative approaches such as the control scheme proposed in [13] could be utilized.

Remark 2. In case a single or more agents N_r are forced to leave the formation, the algorithm can still converges if the rest of robots are informed of the new desired spacing $2\pi/(N - N_r)$. As the leader is randomly elected, a fault affecting the leader is not critical to the proposed startegy.

3   Experiments

3.1   Hardware

A custom made castor wheeled robotic platform equipped with an ultrasonic sensor is utilized in the experiments. The ultrasonic sensor is mounted on a *servo motor* allowing to rotate the device and to measure frontward or backward proximity distance. The analog signal of the ultrasonic sensor is processed by an *Arduino Uno* micro-controller. The state estimation and control algorithm was written with custom python code and run by a Raspberry Pi computer board interfaced through serial communication with the micro-controller to receive sensor data.

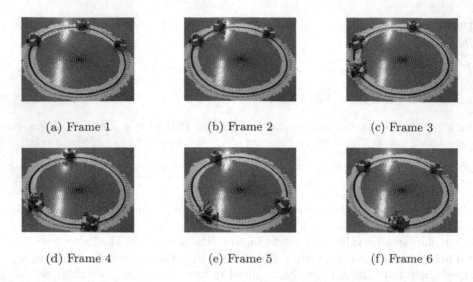

(a) Frame 1   (b) Frame 2   (c) Frame 3

(d) Frame 4   (e) Frame 5   (f) Frame 6

Fig. 1. A sequence of snapshots, sampled at intervals of 200 s, of a sample trial with a team of three robots achieving a circular self-balancing formation.

The *servo-motor* was also set to position the sensor either frontward and backward at three different angular position values spaced by an angle of $7\pi/36$ rad. At each of these positions, eight different measurements were taken and only values between 0.05 m and the circle's diameter were considered and their median computed as the proximity distance to inform the algorithm. Note that, these multiple measurements allowed to compensate for the reduced accuracy of the ultrasonic sensor when the object detected is not directly aligned with the sensor. To fulfill with the constrained imposed to the state estimation and control algorithm, we set the code to discard measurement values greater than a meter yielding to a maximum proximity radius in the output Eq. (3) of $r_v = 1$ m and corresponding to a visibility angle of $\varphi_v = 1.59$ rad.

The robots were programmed to move on a circle identified by a narrow black adhesive tape on top of a wider white adhesive tape. Custom python scripts implementing an independent PID control algorithm with feedback from encoders mounted on the wheels and from the light sensors was utilized to maintain the robot along the black stripe. The radius of the circle was set to $R = 0.7$ m in order to maintain a blind sensing spot of about $\pi/6$ rad prior to reaching the balanced circular formation. The line following algorithm was observed to result in a zig-zag motion generating additional disturbances to the formation control algorithm.

3.2 Procedure

The experiments were performed with group of three robots. A single robot was set as the leader to move at a constant reference angular speed ω_{ref}. All robots, including the leader, were equipped with the ultrasonic sensor not to detect and estimate proximity distance, but also to avoid possible collision with neighbors. This procedure was necessary in case a robot ahead of the team become faulty or unresponsive. We comment that, when the algorithm is successfully implemented by all robots, collision avoidance is useless as the closest robot will detect the motion of the pursuant and initiate its own motion. Further, as the followers rely solely on proximity distances and do not communicate with closer neighbors, a key feature of the state estimation and control algorithm in (6)–(10) is to be fully decentralized and non-cooperative.

The initial position of the robot was set to ensure that two consecutive robots could not detect each other at the beginning of the experiments. Prior to the experiments, pilot trials were conducted allowing to set the pause time interval duration to 55 s. Given that the duration of each time step in Eq. (1) is 1 s, the number of time steps required to *pause* is $p = 55$. Five experimental trials were performed for a total duration of 25 min each. The trials were video recorded using an overhead camera to obtain a wide complete view of the arena. The motion of each robot was then manually extracted from the video frames using a protractor superimposed on top of each picture frames (see Fig. 1).

3.3 Control Parameters

We set the reference angular speed to $\omega_{\mathrm{ref}} = 0.1$ rad/s and we select the control parameter c such that the ε-balanced formation is achieved by the team of robots with a maximum formation error $\varepsilon = 0.4$ rad. Given the trade-off between convergence speed and accuracy (see Remark 1), we select $c = 0.2$ rad/s which is compatible with the maximum formation error above while allowing to fulfill the hypothesis of Propositions 1 and 2. This can be verified for each pair of consecutive agents i and j by evaluating $\varphi_{ij}(0) \in [0.84, 2.02[\subset [\min\{4\varphi_{\max} + 2\omega_{\mathrm{ref}} + 2c, \varphi_v\}, \pi] = [0.68, \pi]^5$, and $\omega_{\mathrm{ref}} + c = 0.3$ rad $< \varphi_v = 1.59$ rad.

4 Results

4.1 Proximity Distances

Table 2 in Appendix presents the raw data estimates of the phase angles measure with the ultrasonic sensors in consecutive iterations after p time steps each. The presence of missing values denoted "n.a." in the Table indicates that the ultrasound sensor often does not return meaningful measurements. In particular, at steady state, measurements might not be available as the desired spacing distance is set at $2\pi/N = 2\pi/3 \simeq 2.09$ rad, much larger than the *visibility cone* that corresponds to a threshold value of $\varphi_v = 1.59$ rad. In the Table, the existence of a measurement value for the backward sensor reading value of robot 1 also indicates that the sensor might also return false readings due to occasional

---

[5] Note that $\varphi_{\max} = 4.2 \times 10^{-2}$ rad since $\delta_{\max} = 0.02$ m. The relationship between $\varphi_{\max}$ and $\delta_{\max}$ is given below Eq. (5).

obstructions of the signal by the operator conducting the experiments. Further, the multiplicity of sound wave emitted by several robots might overlap as the micro-controller cannot differentiate them. As such the control scheme has to deal with additional sources of uncertainties.

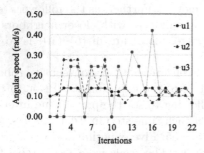 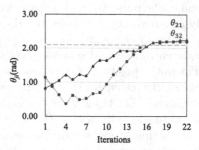

Fig. 2. Evolution of the control law u_i in (7a)–(7b) (left panel) and relative phase angle difference between robots 2 and 1 and between robots 3 and 2 (red panel) in consecutive iterations (every p time steps) for an exemplary trial. The dash-dotted horizontal line in the right panel corresponds to the desired spacing phase angle $2\pi/3$ (Color figure online)

4.2   Control Actuation

The left panel of Fig. 2 depicts the time trace of the control output u_1, u_2, and u_3 observed for each robot. Robots 2 and 3 are able in a few iterations to identify the presence of their closest pursuant by setting their values to 0.3 rad/s in order to space accordingly. We note that the values observed are often different from the control input which should be either 0 initially, then 0.3 rad/s when the motion is initiated, and 0.1 rad/s when the desired spacing is achieved. Note that the large fluctuations of the observed angular speed are explained by several factors including the zig-zagging line following motion, slip, friction, or inaccurate encoder feedback.

Table 1. Convergence rate measured by the number of iterations for robot between robots i to achieve the desired spacing (k_i^c/p), steady-state relative phase angle ($\bar{\Theta}_{ij}$) and maximum formation error ($|\bar{\Theta}_{ij} - \frac{2\pi}{3}|$) between robots i and j. "se" is the standard error.

Measure	k_i^c/p			$\Theta_{i,i-1}$ (rad)		$|\bar{\Theta}_{i,i-1} - \frac{2\pi}{3}|$
Statistics	min	mode	max	mean (rad)	se (rad)	max (rad)
Robot 2	6	7	7	2.02	0.15	0.23
Robot 3	11	11	14	2.20	0.24	0.29

4.3   Convergence

Table 1 presents relevant data of the self-balancing convergence including the number of iterations implemented by a follower robot to reach the desired spacing distance, the averaged relative phase distance between robots, and the maximum formation error observed. The Table shows that the first follower achieves the desired spacing in about 7 iterations while the second robot needs about 11 iterations to converge. The estimated balancing error is always less than 0.30 rad corresponding to the maximum angle value $\omega_{\text{ref}} + c$ spanned by the robot in a single step. This value is also less than the predicted maximum error value 0.40 rad from Proposition 2. In addition, each follower is observed to identify their closest pursuant in a single iteration. Note that the upper bound predicted by (11) is $k_2/p = 3$ and $k_3/p = 10$ for robot 2 and robot 3 respectively.

The right·panel of Fig. 2 depicts the time trace of the relative phase angle difference between robot 2 and robot 1 and between robots 3 and robot 2. In the figure, as time evolves, both values tend to converges toward the desired spacing value $2\pi/N = 2\pi/3 = 2.09$ rad in about 14 iterations. Note that, as discussed in Remark 1, by further reducing the value of the control parameter c, one might further increase the accuracy of the state estimation and control scheme.

5   Conclusion

A non-cooperative and fully decentralized state estimation and control scheme has been experimentally tested. Inspired by the behavior of insects, which stop to enhance the effectiveness of their next move, the scheme is implemented in an *pause-and-go* fashion on a low cost robotic platform consisting of three cantor wheeled robots. It relies on uncertain and intermittent proximity distances measured by an ultrasonic sensor. The estimation and control strategy is capable to autonomously space the robots' along the circle even in case the sensor range is shorter than the desired spacing. By means of a suitable selection of the main control parameters, it is possible to drive the formation error below a desired threshold value and to regulate the trade-off between accuracy and convergence speed. Further, we illustrated that a low-cost implementation of this strategy is robust enough to handle occasional false readings and inaccuracies of the ultrasound sensor. These promising results showing robustness to noise and uncertainties make the proposed approach particularly suitable for applications such as distributed sensor placements or coverage control problems [15]. The proposed state pause-and-go implementation well fits low costs micro-robotic applications that are not time-critical but require limited payload and accuracy. Within formation control problems, the proposed formulation on the circle is relevant to problems such as perimeter surveillance [17] and source seeking [6], and can be possibly extended to more complex shapes assuming they can be approximated with Jordan curves [14]. Depending on the application and on the available budget, different kind of sensors can be employed, see e.g. the discussion in [24] for alternative sensor selections.

Acknowledgments. The authors are grateful to Samuel Coyle and Kevin Lee for contributing during the Summer Program for Undergraduate Research (CU SPUR 2018) in preliminary works to design and fabricate the robotic platform.

Author contributions. V.M., P.D., and F.L.I. designed the study, J.S.C. conducted the experiments on ground robots, V.M., P.D., F.L.I., and J.S.C. performed the analysis, V.M., and P.D. wrote the manuscript, with contributions from all authors.

6   Appendix

Table 2. Proximity distances in radiant (rad) measured frontward and backward by the ultrasonic sensor in an exemplary trial. Note that in case of no measurement or no object detected within the sensing range, the estimated proximity distance is set to "n.a." as in (3). False readings are inside a box.

Iteration	Back 1	Front 1	Back 2	Front 2	Back 3	Front 3
1	n.a.	1.13	n.a.	n.a.	n.a.	n.a.
2	n.a.	n.a.	0.87	2.42	2.12	n.a.
3	n.a.	n.a.	1.61	n.a.	1.34	n.a.
4	n.a.	1.43	n.a.	1.74	0.79	n.a.
5	n.a.	n.a.	n.a.	1.03	1.30	n.a.
6	1.30	n.a.	n.a.	1.06	0.93	n.a.
7	n.a.	n.a.	n.a.	1.11	0.76	n.a.
8	0.85	n.a.	1.29	n.a.	0.84	2.09
9	n.a.	n.a.	n.a.	1.03	1.16	2.38
10	n.a.	n.a.	n.a.	n.a.	1.42	n.a.

Algorithm 1. Implementation of the state estimation in (9). $\lambda(k)$ refers to the number of intervals in $J(k) = -\Upsilon_{ij}(kp) \cup \Upsilon_{ij}(kp)$ and the subroutine Evaluate is defined in Algorithm 2.

1: **procedure** INITIALIZATION ($k = 0$, $k_i < 0$)   ▷ Initially set time $k = 0$ and $k_i < 0$
2:     $J_l(k) = \Upsilon_{ij}(k) \cup -\Upsilon_{ij}(k)$ ▷ Evaluate $J_l(0)$
3:     **if** $\Theta_{ij}(0) \leq \varphi_v$ **then**   ▷ i can detect j
4:         $k_i = 0$ ▷ set $k_i = 0$
5:     **end if**
6:     **while** width($H(kp)$) $\geq \delta$ **do**   ▷ δ is to be defined
7:         $[\lambda(k), J_l(k)] \leftarrow$ Evaluate$\{J_l(k-1) \cap \Gamma_{ij}(kp|(k-1)p)\}$
8:         **if** $d_{ij}(kp) \neq$ n.a. **then**   ▷ A value is returned
9:             **if** $k_i < 0$ **then**
10:                 $k_i = kp$
11:             **end if**
12:         **else**   ▷ No value is returned
13:             **if** $\varphi_v \leq \frac{\pi}{3}$ **then**
14:                 $\lambda(k) = 2$ ▷ $\Gamma_{ij}(kp|kp)$ has two intervals
15:             **else**
16:                 $\lambda(k) \leq 1$ ▷ $\Gamma_{ij}(kp|kp)$ is either the empty set or a single interval
17:             **end if**
18:         **end if**
19:         $k = k + 1$
20:         $J_l(k) = -\Upsilon_{ij}(kp) \cup \Upsilon_{ij}(kp)$
21:         width $(H(kp)) = \max_l J_l(k) - \min_l J_l(k)$
22:     **end while**
23:     **return** $J_l(k|k)$
24: **end procedure**

Algorithm 2. Subroutine of Algorithm 1.

1: **procedure**
2:     Evaluate$\{J_l(k|k-1) \cap \Gamma_{ij}(kp|(k-1)p)\}$
3:     $\lambda(k) = 0$
4:     $\ddot{\varphi}_{ij}(kp) = 2\arcsin(\ddot{d}_{ij}(kp)/2R)$ ▷ Sensor measurements
5:     $\Upsilon_{ij}(kp) := [\max\{\tilde{\varphi}_{ij}(kp) - \varphi_{\max}, 0\}, \min\{\tilde{\varphi}_{ij}(kp) + \varphi_{\max}, \varphi_v\}]$ ▷ Update
6:     **for** $l = 1:N^l$ **do**
7:         **if** $J_l(k|k-1) \cap \Gamma_{ij}(kp|(k-1)p)$ is a single interval **then**
8:             $\lambda(k) = \lambda(k) + 1$
9:         **else if** $J_l(k|k-1) \cap \Gamma_{ij}(kp|(k-1)p)$ is the union of two intervals **then**
10:             $\lambda(k) = \lambda(k) + 2$
11:         **end if**
12:     **end for**
13:     **return** $[\lambda(k), J_l(k|k-1) \cap \Gamma_{ij}(kp|(k-1)p)]$
14: **end procedure**

References

1. Ahmed, N., Cortes, J., Martinez, S.: Distributed control and estimation of robotic vehicle networks: overview of the special issue. IEEE Control Syst. Mag. **36**(2), 36–40 (2016)
2. Ariel, G., Ophir, Y., Levi, S., Ben-Jacob, E., Ayali, A.: Individual pause-and-go motion is instrumental to the formation and maintenance of swarms of marching locust nymphs. PLoS ONE **9**(7), e101636 (2014)
3. Bernard, M., Kondak, K., Maza, I., Ollero, A.: Autonomous transportation and deployment with aerial robots for search and rescue missions. J. Field Robot. **28**(6), 914–931 (2011)
4. Bopardikar, S.D., Englot, B., Speranzon, A.: Robust belief roadmap: planning under uncertain and intermittent sensing. In: 2014 IEEE International Conference on Robotics and Automation (ICRA), pp. 6122–6129. IEEE (2014)
5. Borenstein, J., Feng, L.: Measurement and correction of systematic odometry errors in mobile robots. IEEE Trans. Robot. Autom. **12**(6), 869–880 (1996)
6. Briñón-Arranz, L., Schenato, L., Seuret, A.: Distributed source seeking via a circular formation of agents under communication constraints. IEEE Trans. Control Netw. Syst. **3**(2), 104–115 (2015)
7. Chen, J., Sun, D.: Resource constrained multirobot task allocation based on leader-follower coalition methodology. Int. J. Robot. Res. **30**(12), 1423–1434 (2011)
8. Chen, Y.Q., Wang, Z.: Formation control: a review and a new consideration. In: 2005 IEEE/RSJ International Conference on Intelligent Robots and Systems, pp. 3181–3186. IEEE (2005)
9. Cho, B.S., Moon, W., Seo, W.J., Baek, K.R.: A dead reckoning localization system for mobile robots using inertial sensors and wheel revolution encoding. J. Mech. Sci. Technol. **25**(11), 2907–2917 (2011). https://doi.org/10.1007/s12206-011-0805-1
10. Couzin, I.D., Krause, J., Franks, N.R., Levin, S.A.: Effective leadership and decision-making in animal groups on the move. Nature **433**(7025), 513–516 (2005)
11. D'Amico, S., Montenbruck, O.: Differential GPS: an enabling technology for formation flying satellites. In: Sandau, R., Roeser, H.P., Valenzuela, A. (eds.) Small Satellite Missions for Earth Observation, pp. 457–465. Springer, Heidelberg (2010). https://doi.org/10.1007/978-3-642-03501-2_43
12. DeLellis, P., Garofalo, F., Iudice, F.L., Mancini, G.: Balancing cyclic pursuit using proximity sensors with limited range. IFAC Proc. Vol. **47**(3), 5784–5789 (2014)
13. DeLellis, P., Garofalo, F., Iudice, F.L., Mancini, G.: State estimation of heterogeneous oscillators by means of proximity measurements. Automatica **51**, 378–384 (2015)
14. DeLellis, P., Garofalo, F., Lo Iudice, F., Mancini, G.: Decentralised coordination of a multi-agent system based on intermittent data. Int. J. Control **88**(8), 1523–1532 (2015)
15. Dou, L., Song, C., Wang, X., Liu, L., Feng, G.: Coverage control for heterogeneous mobile sensor networks subject to measurement errors. IEEE Trans. Autom. Control **63**(10), 3479–3486 (2018)
16. Dudek, G., Jenkin, M.: Computational Principles of Mobile Robotics. Cambridge University Press, Cambridge (2010)
17. Elmaliach, Y., Agmon, N., Kaminka, G.A.: Multi-robot area patrol under frequency constraints. Ann. Math. Artif. Intell. **57**(3–4), 293–320 (2009). https://doi.org/10.1007/s10472-010-9193-y

18. Erdelj, M., Król, M., Natalizio, E.: Wireless sensor networks and multi-UAV systems for natural disaster management. Comput. Netw. **124**, 72–86 (2017)
19. Jetto, L., Longhi, S., Venturini, G.: Development and experimental validation of an adaptive extended Kalman filter for the localization of mobile robots. IEEE Trans. Robot. Autom. **15**(2), 219–229 (1999)
20. Le Bars, F., Sliwka, J., Jaulin, L., Reynet, O.: Set-membership state estimation with fleeting data. Automatica **48**(2), 381–387 (2012)
21. Lo Iudice, F., Acosta, J.A., Garofalo, F., DeLellis, P.: Estimation and control of oscillators through short-range noisy proximity measurements. Automatica **113**, 108752-1–108752-8 (2020)
22. Luo, R.C., Yih, C.C., Su, K.L.: Multisensor fusion and integration: approaches, applications, and future research directions. IEEE Sens. J. **2**(2), 107–119 (2002)
23. Mariottini, G.L., Morbidi, F., Prattichizzo, D., Pappas, G.J., Daniilidis, K.: Leader-follower formations: uncalibrated vision-based localization and control. In: Proceedings of the 2007 IEEE International Conference on Robotics and Automation, pp. 2403–2408. IEEE (2007)
24. Mwaffo, V., Curry, J.S., Iudice, F.L., De Lellis, P.: Pause-and-go self-balancing formation control of autonomous vehicles using vision and ultrasound sensors. IEEE Trans. Control Syst. Technol. **29**(6), 2299–2311 (2021)
25. Oh, K.K., Park, M.C., Ahn, H.S.: A survey of multi-agent formation control. Automatica **53**, 424–440 (2015)
26. Stroupe, A.W., Martin, M.C., Balch, T.: Distributed sensor fusion for object position estimation by multi-robot systems. In: Proceedings of the 2001 ICRA. IEEE International Conference on Robotics and Automation (Cat. No. 01CH37164), vol. 2, pp. 1092–1098. IEEE (2001)
27. Zabala, F., Polidoro, P., Robie, A., Branson, K., Perona, P., Dickinson, M.H.: A simple strategy for detecting moving objects during locomotion revealed by animal-robot interactions. Curr. Biol. **22**(14), 1344–1350 (2012)
28. Zhang, Q., Lapierre, L., Xiang, X.: Distributed control of coordinated path tracking for networked nonholonomic mobile vehicles. IEEE Trans. Ind. Inf. **9**(1), 472–484 (2012)

Bio-inspired Information
and Communication

Evolution of Vesicle Release Mechanisms in Neuro-Spike Communication

Yu Wenlong and Lin Lin[✉]

College of Electronics and Information Engineering, Tongji University,
Shanghai 201804, China
fxlinlin@tongji.edu.cn

Abstract. Neuro-spike communication has become a hot topic and has been investigated extensively in recent years. The vesicle release process is the main part of neuro-spike communication, which directly determines the accuracy of information transmission. Currently, single vesicle release (SVR) model is used to investigate the process of vesicle release, but there is few research on multi-vesicle release (MVR) model. In this paper, a pool-based MVR model is presented, and the influence of data rate and several system parameters on bit error rate (BER) performance of the model is simulated. In addition, the BER performance of SVR model and MVR model under the same conditions is compared. The advantages and disadvantages of the two models are analyzed.

Keywords: Molecular communication · Neuro-spike communication · Nanonetworks · Vesicle release process · Bit error rate

1 Introduction

As the most promising candidate for reliable communications at nano- or micro-scale, molecular communication has attracted much attention in recent years. Great progress has been made in the research on molecular communication. Many kinds of molecular communication systems have been investigated such as molecular communication via diffusion [11–13,15,25,32], microtubule-based communication [4,24], pheromone signaling [8], and bacteria-based communication [9].

Neuro-spike communication is a kind of communication method that includes electrical process and molecular communication process. The performance of

This work is supported in part by National Natural Science Foundation, China (61971314), in part by Science and Technology Commission of Shanghai Municipality (19510744900), and in part by Sino-German Center of Intelligent Systems, Tongji University.

© ICST Institute for Computer Sciences, Social Informatics and Telecommunications Engineering 2021
Published by Springer Nature Switzerland AG 2021. All Rights Reserved
T. Nakano (Ed.): BICT 2021, LNICST 403, pp. 105–116, 2021.
https://doi.org/10.1007/978-3-030-92163-7_9

neuro-spike communications is quite good in terms of reliability, speed and robustness [5]. The analysis of this communication paradigm is beneficial to exploit in the artificial neural systems where nanomachines are linked to neurons to treat the neurodegenerative diseases [23]. Therefore, research on neuro-spike communication has become a hot pot.

Many works have been done in the neuro-spike communication field. Several communication channel models have been introduced for every biological processes of this communication system [2,26,31]. To characterize the fundamental properties of neuro-spike communication, a physical channel model was proposed in [2]. Reference [20] proposed a synaptic model which shows that redundancy of synapses increases the information transmission efficiency. An alternative representation of the neuron-to-neuron communication method was proposed in [31] where the biological processes are modeled by their frequency responses. The vital events during the synaptic transmission were investigated in [21]. As the major part of neuro-spike communication, vesicle release process has received much attention.

Vesicle release is the process that the spike propagating to the end of the axon prompts the pre-synaptic terminals to release vesicles into gaps. Many vesicle release models have been investigated. In [31], a model which has the fixed vesicle release probability is utilized. There were finite state markov channels model used in [2] and pool-based release model used in [19,22]. However, the pool-based model is not realistic, where the number of available vesicles are overestimated since the refill rate is assumed to be proportional to the number of reserve vesicles. A realistic pool-based model for vesicle release and replenishment was proposed in [27], in which channel characteristics of neuro-spike communication systems have been investigated. These models only study the single vesicle release (SVR) process, and there is limited research on multi-vesicle release (MVR) models. But in fact, the SVR was originally thought to liberate at most one vesicle for each spike, rendering synaptic transmission unreliable. MVR, which was initially identified at specialized synapses but is now known to be common throughout the brain [28], represents a simple mechanism to overcome the intrinsic unreliability of synaptic transmission.

This paper presents a pool-based MVR and replenishment model. The BER performance of this model is investigated under different conditions comparing with that of SVR model. The release process of the two model have been introduced in [22]. Only a vesicle is released in response to a spike that inhibits the release of other vesicles in SVR process, while the number of vesicles released is random when a spike arrives in MVR process. The vesicle replenishment process is described in [27]. Besides, the influence of parameters of the two vesicle release models and data transmission rate on error probability is simulated. We compare the BER performance of the two models under different conditions and analyze the reasons in detail. The main contributions of this paper lie in several aspects:

– A realistic MVR model is presented to investigate the BER performance.
– The influence of parameters involved in the vesicle release process and data
  rate on BER performance is investigated in detail.

The remainder of this paper is organized as follows. Section 2 introduces
the system model of neuro-spike communication. In Sect. 3, the vesicle release
process of two vesicle release models is analyzed. The BER of two models is simulated under different conditions by MATLAB in Sect. 4. The influence of different
parameters on BER is investigated and the reasons are analyzed. Finally, Sect. 5
concludes this paper.

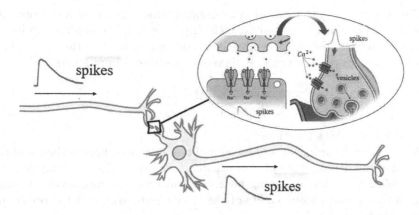

Fig. 1. The neuro-spike communication process. When a spike arrives at the end of the
axon, the pre-synaptic terminals will release vesicles into gaps. The neurotransmitters
released by the vesicles arrive at postsynaptic receptor after diffusion, and a new spike
is formed on the next neuron.

2   System Model

The human nervous system contains billions of interconnected neurons that communicate with each other through synapses. The neuron is in a resting manner
and is polarized with an intracellular potential about -95 to -65 mV when no
signal is transmitted via the nervous system [10]. The nervous system transmits
signals by electrically charged ion flows of potassium (K^+), sodium (Na^+), chlo-
ride (Cl^-). To transmit signals, neuron membrane potential can be changed by
ion exchange between inside and outside of the neuron, which enter the neuron or exit from that via the ion channels located on the soma and dendrites
(cation and anion channels) and on the nodes of Ranvier (Sodium and potassium channels). The potential increases high enough about 20 mV, the neuron
will be excited and the membrane will be depolarized. Then, the firing will hap-
pen. When a neuron fires, an spike about 90 mV at a time period of 1 ms will be
generated in the neuron. When a spike reaches an axon terminal, the depolariza-
tion leads to opening the calcium channels and causes an influx of calcium ions

(Ca^{2+}) into the pre-synaptic neuron [16], which leads to the release of neuro-transmitters into the synapse gap. Neurotransmitters bind to the receptor of the postsynaptic neuron and make the cation or anion channel open by changing of permeability features of the neuron as shown in Fig. 1. Opening cation channels conduct positively charged ions into the neuron, and thus, increases its potential and leads to a spike firing.

Neuro-spike communication model is a hybrid model that involves both molecular communication in synaptic transmission part, and electrical trans-mission of action potentials in the axonal pathway. The traditional neuro-spike communication model is generally divided into three phases based on the bio-logical process [1].

The first phase is the axonal transmission, that the spikes are propagated along the axon. When a spike arrives, the input symbol is considered to be "1". If no spike arrives, the input symbol is considered to be "0". Hence, the input $s(t)$ of pre-synaptic neuron can be viewed as a series of impulse signals:

$$s(t) = \sum_i \delta(t - t_i), \tag{1}$$

where $\delta(\cdot)$ is the delta function.

When the spike propagates to the end of the axon and changes the membrane potential of the neuron, the influx of calcium ions causes the pre-synaptic termi-nals to release vesicles containing neurotransmitters into gaps among neurons, which is the second phase. The pool-based vesicle release model is considered, which is shown in Fig. 2.

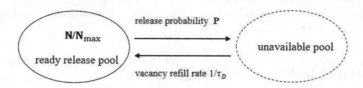

Fig. 2. Pool-based vesicle release model.

Readily releasable vesicles (RRVs) are stored in a pool with an upper limit of N_{max} called ready pool (RP), while the others are contained in a larger unavailable pool away from the pre-synaptic. Once a vesicle is released, there is a vacancy in RP and the vesicles farther away from the pre-synaptic are refilled into RP until the number of vesicles reaches N_{max}. It is assumed that the number of ready to be docked in neurons is much higher than N_{max} and the recovery of each vacancy is independent of others. So the recovery of a vacancy can be modeled by the first event from a Passion process [27]. Therefore, in a symbol interval, the probability of one vacancy replenishment P_{rep} is described as:

$$P_{rep} = 1 - \exp(-\tau_D^{-1} T_s), \tag{2}$$

where τ_D is the mean recovery time of a vacancy [22]. The number of vesicles recovered after Δt obeys the Binomial distribution $B(N_{max} - N, P_{rep})$. The vesicle release process is described by the following two models.

In the SVR model introduced in [27], when input symbol is "1" and the number of RRVs in RP N is larger than 0, a RRV is released as a result. At the same time, it transiently prevents other vesicles from being released [14]. Based on [7], the probability that no RRVs is released per stimulus is $\exp(-a_v N)$ for RP with N RRVs ($0 < N \leq N_{max}$), where a_v is the vesicle fusion rate. Therefore, the release probability of a RRV P_s when there are N RRVs in RP is depicted as

$$P_s = 1 - \exp(-a_v N). \tag{3}$$

Besides, the vesicle fusion rate a_v is related to the number of vesicles N in RP [7] and can be expressed as

$$a_v = k_a \sqrt{N}, \tag{4}$$

where k_a is a positive coefficient. Therefore, (3) can also be described as

$$P_s = 1 - \exp(-k_a N^{\frac{3}{2}}). \tag{5}$$

A pool-based MVR model is presented in this paper. In this model, MVR is allowed and individual vesicles are released independently of each other. Therefore, the release probability of a RRV P_m is expressed as

$$P_m = 1 - \exp(-a_v) = 1 - \exp(-k_a \sqrt{N}). \tag{6}$$

So when input symbol is "1" and the RP contains N RRVs, the number of vesicles released obeys the binomial distribution $D(N, \Gamma_m)$.

Assuming that the symbol interval T_s is larger than the pulse duration, $s(t)$ can be described as

$$s(t) = \sum_{i=1}^{K} N_i a_i \delta(t - (i - 1)T_s), \tag{7}$$

where a_i is the ith input symbol, K is the length of input signal, N_i is the number of neurotransmitters in all vesicles released for the first symbol.

The final phase is that the neurotransmitter spreads to the postsynaptic receptor and binds to it, forming a new spike on the next neuron and completing the transmission of information between neurons. Based on Fick's second law of diffusion which states that at time t, at position x, the molecular concentration $C(x, t)$ can be described by

$$\frac{1}{D} \frac{\partial C(x, t)}{\partial t} = \nabla^2 C(x, t), \tag{8}$$

where D is the diffusion coefficient of the medium, ∇^2 is Laplacian operator. The channel model proposed in [17] is utilized. Particle re-uptake at the pre-synaptic neuron can be modeled as irreversible adsorption to the homogeneous left boundary. The radiating boundary condition modeling pre-synaptic re-uptake is given as

$$D\frac{\partial C(x,t)}{\partial x} = k_r C(x,t), \text{ at } x = 0, \tag{9}$$

where k_r is the re-uptake coefficient. For the right boundary, the radiating boundary needs to be extended to incorporate particle desorption. Particle desorption is modeled as a first-order process depending on the intrinsic desorption rate k_d and on the amount of currently adsorbed particles, as

$$D\frac{\partial C(x,t)}{\partial x} = -k_f C(x,t) - k_d \int_0^t D\frac{\partial C(x,\tau)}{\partial x} d\tau, \text{ at } x = d, \tag{10}$$

where k_f is the effective association coefficient.

Based on (7), the number of neurotransmitters released at $t = 0$ can be modeled with the initial value as

$$C(x,0) = N_1 \delta(x). \tag{11}$$

By finding the $C(x,t)$ that satisfies (8), (9), (10) and (11), the channel impulse response $h(t)$ can be expressed as

$$h(t) = \int_0^t -D\frac{\partial C(x,\tau)}{\partial x}\bigg|_{x=d} d\tau, \tag{12}$$

where d is the length of synaptic gap. So the response of channel $y(t)$ after free diffusion is

$$y(t) = s(t) * h(t), \tag{13}$$

where symbol $*$ indicates the convolution operation.

The energy difference based detection method is used to receive the signal according to [18].

3   Theoretical Analysis

When an impulse arrives at the end of axon, the MVR model may release more vesicles than SVR model under the same conditions. But at the same time, the MVR model with more vacancies may also have a higher number of vesicle replenishment. Then, it is difficult to compare vesicle consumption rates between the two models. Therefore, the two release models are analyzed in detail in this section.

Assuming that there are N vesicles in RP at time t ($0 < N < N_{max}$). The vesicles replenishment process of the two vesicle release models is the same according to Sect. 2. For SVR model, the number of vesicles in RP is N_s at $t+T_s$ ($N-1 \leq N_s \leq N_{max}$). The corresponding probability is:

$$Pr(N_s = N - 1) = (1 - P_{rep})^{N_{\max}-N} P_s,$$
$$Pr(N_s = N) = (1 - P_{rep})^{N_{\max}-N}(1 - P_s)$$
$$+ (1 - P_{rep})^{N_{\max}-N-1} P_s,$$
$$\ldots$$
$$Pr(N_s = N + i) = \binom{i}{N_{\max}-N}(1 - P_{rep})^{N_{\max}-N-i} P_{rep}^i (1 - P_s)$$
$$+ \binom{i+1}{N_{\max}-N}(1 - P_{rep})^{N_{\max}-N-i-1} P_{rep}^{i-1} P_s, \tag{14}$$
$$\ldots$$
$$Pr(N_s = N_{\max}) = P_{rep}^{N_{\max}-N}(1 - P_s).$$

For simplicity, P_i is used to express $Pr(N_s = N + i)$ $(-1 \le i \le N_{\max} - N)$. According to the knowledge of probability theory, vesicle release and recovery process are independent of each other. Hence the mean of the variation number of vesicles in RP is described as

$$E_s = \sum_{i=-1}^{N_{\max}-N} i \times P_i = E_{rep1} + E_{rel1}, \tag{15}$$

where E_{rep1} is the mean of vesicle recovery process and E_{rel1} is the mean of vesicle release process. Therefore,

$$E_s = (N_{\max} - N)P_{rep} - P_s. \tag{16}$$

In the same way, the mean of the variation number of vesicles in RP for MVR model is depicted as

$$E_m = (N_{\max} - N)P_{rep} - NP_m. \tag{17}$$

Based on (2)–(6), it can be seen that both E_s and E_m are functions of N. To investigate the properties of these two mean functions, the derivatives of E_s and E_m are obtained, as

$$\frac{\mathrm{d}E_s(N)}{\mathrm{d}N} = -P_{rep} - \frac{3k_a\sqrt{N}}{2}\exp(-k_a N^{3/2}),$$
$$\frac{\mathrm{d}E_m(N)}{\mathrm{d}N} = -P_{rep} - 1 + \exp(-k_a N^{1/2}) - \frac{k_a\sqrt{N}}{2}\exp(-k_a N^{1/2}), \tag{18}$$

where P_{rep}, k_a and N are all larger than 0. From (18), it can be seen that both $\frac{\mathrm{d}E_s(N)}{\mathrm{d}N}$ and $\frac{\mathrm{d}E_m(N)}{\mathrm{d}N}$ are smaller than 0 when $N \in [0, N_{\max}]$. Hence $E_s(N)$ and $E_m(N)$ are both monotonically decreasing functions of N. In addition, based on (16) and (17), we obtain

$$E_s(0) = N_{\max}P_{rep} > 0,$$
$$E_s(N_{\max}) = -1 + \exp(-k_a N_{\max}^{3/2}) < 0,$$
$$E_m(0) = N_{\max}P_{rep} > 0, \tag{19}$$
$$E_m(N_{\max}) = -N_{\max}(1 - \exp(-k_a N_{\max}^{1/2})) < 0.$$

So both $E_s(N)$ and $E_m(N)$ have unique zero point in $(0, N_{\max})$, which can be obtained by solving $E_s(N) = 0$ and $E_m(N) = 0$. x is used to represent $\sqrt{N}$ in (16) and (17) for simplicity, which is expressed as

$$f_s(x) = (N_{\max} - x^2)P_{rep} - 1 + \exp(-k_a x^3) = 0,$$
$$f_m(x) = (N_{\max} - x^2)P_{rep} - x^2(1 - \exp(-k_a x)) = 0, \tag{20}$$

where $x \in (0, \sqrt{N_{\max}})$. After simplification, Eq. (20) can be expressed as

$$x_s^3 + \frac{1 - N_{\max}P_{rep}}{P_{rep}} x_s + \frac{2}{3k_a} = 0,$$

$$x_m^3 - \frac{N_{\max}P_{rep}}{1 + P_{rep}} x_m + \frac{2}{k_a} \frac{N_{\max}P_{rep}}{1 + P_{rep}} = 0, \tag{21}$$

where x_s, x_m are the zero points of $f_s(x) = 0$ and $f_m(x) = 0$, respectively. Based on Cardano formula [30], we can obtain

$$x_s = \left(-\frac{q_s}{2} + \sqrt{\frac{r_s^3}{27} + \frac{q_s^2}{4}}\right)^{1/3} + \left(-\frac{q_s}{2} - \sqrt{\frac{r_s^3}{27} + \frac{q_s^2}{4}}\right)^{1/3},$$

$$x_m = \left(-\frac{q_m}{2} + \sqrt{\frac{r_m^3}{27} + \frac{q_m^2}{4}}\right)^{1/3} + \left(-\frac{q_m}{2} - \sqrt{\frac{r_m^3}{27} + \frac{q_m^2}{4}}\right)^{1/3}, \tag{22}$$

where,

$$q_s = \frac{1 - N_{\max}P_{rep}}{P_{rep}},$$

$$r_s = \frac{2}{3k_a},$$

$$q_m = -\frac{N_{\max}P_{rep}}{1 + P_{rep}}, \tag{23}$$

$$r_m = \frac{2}{k_a} \frac{N_{\max}P_{rep}}{1 + P_{rep}}.$$

Besides, among x_s and x_m which are obtained by solving (22), only those satisfying $x_s, x_m \in (0, \sqrt{N_{\max}})$ are wanted.

Their physical meaning can be described: When the time is long enough, the number of vesicles in RP of SVR model will converge to N_{s0} which is equal to x_s^2. As long as $N < N_{s0}$, $E_s(N) > 0$ which means that the vesicle refill rate is larger than release rate. In the same way, the vesicle refill rate is lower than release rate as $N > N_{s0}$. N_{m0} equal to x_m^2 is the convergence value of the vesicle number in RP after a long enough period for MVR model.

4   Simulation Results

Considering the influence of the duration of a spike on vesicle release, the time is discretized into windows with equal width Δt. The window size is selected to ensure that at most one spike exists in each time window. In the simulations of this paper, the width of time window is set equal to the spike duration. Based on [3], $\Delta t = \Delta t_s = 4$ ms. Hence, the symbol interval T_s is described as

$$T_s = k_s \Delta t, \tag{24}$$

where k_s is a positive coefficient and set to be a integer. The data rate is defined as $1/T_s$.

What's more, the mean recovery time τ_D is related to the capacity of the RP [27], as

$$\tau_D = \frac{k_D}{N_{\max}}, \tag{25}$$

where k_D is also a positive coefficient. Default values for simulation parameters are given below: $T_s = 16$ ms, bit length $K = 10^4$, $a_v = 0.06\sqrt{N}$ [7], $N_{\max} = 10$ [29], $\tau_D = \frac{600}{N_{\max}}$ ms [6].

The influence of mean recovery time τ_D on BER is simulated. In each simulation, τ_D changes from 20 to 100 ms with the step of 10 ms. BER for systems utilizing SVR and MVR models with different τ_D is shown in Fig. 3(a).

From Fig. 3(a), it can be seen that the BER of two system models increases as τ_D increases. When τ_D is large, the vesicle replenishment probability is low based on (2) and it will limit the vesicle replenishment rate. Therefore, there are fewer RRVs than the model with smaller τ_D. According to (5), the low number of RRVs will decrease the probability of vesicle release, which will increase the BER.

Then, the influence of capacity of RP $N_{\max}$ on BER is simulated. In each simulation, $N_{\max}$ changes from 5 to 45 with the step of 5. The relationships between BER of the two systems and $N_{\max}$ are simulated and depicted in Fig. 3(b).

Figure 3(b) indicates that larger $N_{\max}$ will cause lower BER of communication model. Even when $N_{\max}$ is large enough, the BER of both systems decreases to 0. When $N_{\max}$ is large, the average recovery time of a vacancy is shorter according to (25). Therefore, there are more vesicles in RP and the vesicle release probability is larger, which leads to a lower BER.

Subsequently, the influence of data rate on BER is simulated. In each simulation, Symbol interval changes from 4 to 36 ms with the step of 4 ms. The relationships between BER of different systems and data rate are simulated and depicted in Fig. 3(c).

Based on the relationship between symbol interval and data rate, when the symbol interval increases, data rate will decrease. As can be seen from Fig. 3(c), the BER gradually decreases with the decrease of data rate. The reason is, the consumption rate of RRVs will be higher when the data rate is high, resulting in smaller RRVs in RV. Hence, the BER will decrease with the decrease of data rate.

The influence of fusion rate a_v is simulated finally. k_a changes from 0.01 to 0.46 with the step of 0.05. The BER of the two models with different a_v is calculated and shown in the Fig. 3(d).

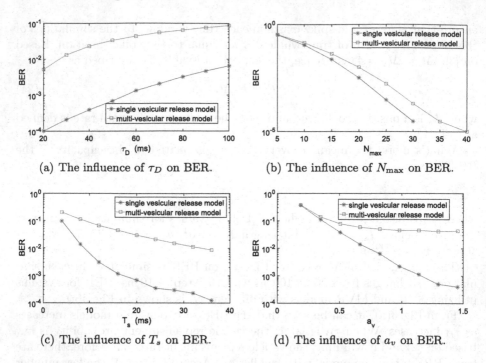

(a) The influence of τ_D on BER.   (b) The influence of N_{max} on BER.

(c) The influence of T_s on BER.   (d) The influence of a_v on BER.

Fig. 3. The influence of parameters on the BER performance of two vesicle release models.

In Fig. 3(d), the relationship between vesicle fusion rate a_v and BER is simulated and shown in Fig. 3(d). From (3) and (6), the increase of a_v will directly reduce the probability of vesicle release, resulting in more error detection cases.

In addition, it can be seen from the Fig. 3 that the BER of SVR model is always lower than the MVR model under the same conditions. This is because the vesicle depletion rate of the MVR model is much higher than that of the SVR model. Therefore, the number of vesicles in RP remains small and the possibility of vesicle depletion is much higher than that in the SVR model. This indicates that limiting the number of vesicles released will increase the transmission accuracy and make more effective use of limited resources to some extent. Also, the BER of SVR model is all more sensitive to the change of parameter than MVR model for Fig. 3. This is because the release of multiple vesicles greatly reduces the contingency of vesicle release and increases the stability of the model.

5   Conclusion

In this paper, a pool-based MVR model is proposed. The influence of data rate, mean vesicle recovery time, capacity of RP and vesicle fusion rate on BER performance of this model is simulated comparing with that of SVR model. From the simulation results, the BER of two models will increase as data rate and mean recovery time of a vacancy increase, while it will decrease when the capacity of RP and vesicle fusion rate become larger. All the parameters change BER by influencing vesicle release probability. Compared with SVR model, MVR model has higher BER under the same conditions but has better stability. That is, all these parameters have greater influence on the BER performance of the SVR model.

References

1. Aghababaiyan, K., Maham, B.: Error probability analysis of neuro-spike communication channel. In: Proceedings of the IEEE Symposium on Computers and Communications (ISCC), pp. 932–937. IEEE (2017)
2. Balevi, E., Akan, O.B.: A physical channel model for nanoscale neuro-spike communications. IEEE Trans. Commun. **61**(3), 1178–1187 (2013)
3. Bear, M.F., Connors, B.W., Paradiso, M.A.: Neuroscience: Exploring the Brain (2007)
4. Darchini, K., Alfa, A.S.: Molecular communication via microtubules and physical contact in nanonetworks: a survey. Nano Commun. Netw. **4**(2), 73–85 (2013)
5. Dayan, P., Abbott, L.: Neuroscience: Computational and Mathematical Modeling of Neural Systems. MIT Press, Cambridge (2001)
6. De La Rocha, J., Parga, N.: Short-term synaptic depression causes a non-monotonic response to correlated stimuli. J. Neurosci. **25**(37), 8416–8431 (2005)
7. Dobrunz, L.E., Stevens, C.F.: Heterogeneity of release probability, facilitation, and depletion at central synapses. Neuron **18**(6), 995 1008 (1997)
8. Giné, L.P., Akyildiz, I.F.: Molecular communication options for long range nanonetworks. Comput. Netw. **53**(16), 2753–2766 (2009)
9. Gregori, M., Akyildiz, I.F.: A new nanonetwork architecture using flagellated bacteria and catalytic nanomotors. IEEE J. Sel. Areas Commun. **28**(4), 612–619 (2010)
10. Hosseini, M., Ghazizadeh, R., Farhadi, H.: A model for electro-chemical neural communication. In: Alam, M.M., Hämäläinen, M., Mucchi, L., Niazi, I.K., Le Moullec, Y. (eds.) BODYNETS 2020. LNICST, vol. 330, pp. 137 150. Springer, Cham (2020). https://doi.org/10.1007/978-3-030-64991-3_10
11. Huang, L., Lin, L., Liu, F., et al.: Clock synchronization for mobile molecular communication systems. IEEE Trans. Nanobiosci. **20**(4), 406–415 (2021)
12. Huang, S., Lin, L., Guo, W., et al.: Initial distance estimation and signal detection for diffusive mobile molecular communication. IEEE Trans. Nanobiosci. **19**(3), 422–433 (2020)
13. Huang, S., Lin, L., Xu, J., Guo, W., et al.: Molecular communication via subdiffusion with a spherical absorbing receiver. IEEE Wirel. Commun. Lett. **9**(10), 1682–1686 (2020)
14. Korn, H., Mallet, A., Triller, A., et al.: Transmission at a central inhibitory synapse. II. Quantal description of release, with a physical correlate for binomial n. J. Neurophysiol. **48**(3), 679–707 (1982)
15. Kuran, M.Ş, Yilmaz, H.B., Tugcu, T., et al.: Energy model for communication via diffusion in nanonetworks. Nano Commun. Netw. **1**(2), 86–95 (2010)
16. Lai, K.O., Ip, N.Y.: Central synapse and neuromuscular junction: same players, different roles. Trends Genet. **19**(7), 395–402 (2003)

17. Lotter, S., Ahmadzadeh, A., Schober, R.: Channel modeling for synaptic molecular communication with re-uptake and reversible receptor binding. In: Proceedings of the IEEE International Conference on Communications (ICC), pp. 1–7. IEEE (2020)
18. Luo, C., Wu, X., Lin, L., et al.: Non-coherent signal detection technique for mobile molecular communication at high data rates. In: Proceedings of the IEEE GLOBE-COM, pp. 1–6. IEEE (2019)
19. Malak, D., Akan, O.B.: A communication theoretical analysis of synaptic multiple-access channel in hippocampal-cortical neurons. IEEE Trans. Commun. **61**(6), 2457–2467 (2013)
20. Manwani, A.: Information-theoretic analysis of neuronal communication. Ph.D. thesis, California Institute of Technology (2000)
21. Manwani, A., Koch, C.: Detecting and estimating signals over noisy and unreliable synapses: information-theoretic analysis. Neural Comput. **13**(1), 1–33 (2001)
22. Matveev, V., Wang, X.J.: Implications of all-or-none synaptic transmission and short-term depression beyond vesicle depletion: a computational study. J. Neurosci. **20**(4), 1575–1588 (2000)
23. Mesiti, F., Balasingham, I.: Nanomachine-to-neuron communication interfaces for neuronal stimulation at nanoscale. IEEE J. Sel. Areas Commun. **31**(12), 695–704 (2013)
24. Moore, M., Enomoto, A., Nakano, T., et al.: A design of a molecular communication system for nanomachines using molecular motors. In: Proceedings of the 4th Annual IEEE International Conference on Pervasive Computing and Communications Workshops (PERCOMW 2006), pp. 6–pp. IEEE (2006)
25. Nakano, T., Eckford, A.W., Haraguchi, T.: Molecular Communication. Cambridge University Press, Cambridge (2013)
26. Ramezani, H., Akan, O.B.: A communication theoretical modeling of axonal propagation in hippocampal pyramidal neurons. IEEE Trans. Nanobiosci. **16**(4), 248–256 (2017)
27. Ramezani, H., Akan, O.B.: Information capacity of vesicle release in neuro-spike communication. IEEE Commun. Lett. **22**(1), 41–44 (2017)
28. Rudolph, S., Tsai, M.C., von Gersdorff, H., et al.: The ubiquitous nature of multivesicular release. Trends Neurosci. **38**(7), 428–438 (2015)
29. Schikorski, T., Stevens, C.F.: Quantitative ultrastructural analysis of hippocampal excitatory synapses. J. Neurosci. **17**(15), 5858–5867 (1997)
30. Tokunaga, H.: Triple coverings of algebraic surfaces according to the Cardano formula. J. Math. Kyoto Univ. **31**(2), 359–375 (1991)
31. Veletić, M., Floor, P.A., Babić, Z., et al.: Peer-to-peer communication in neuronal nano-network. IEEE Trans. Commun. **64**(3), 1153–1166 (2016)
32. Zheng, R., Lin, L., Yan, H.: A noise suppression filter for molecular communication via diffusion. IEEE Wirel. Commun. Lett. **10**(3), 589–593 (2021)

Digestive System Dynamics in Molecular Communication Perspectives

Dixon Vimalajeewa$^{(\boxtimes)}$ and Sasitharan Balasubramaniam

Walton Institute, Waterford Institute of Technology, Waterford, Ireland
{dixon.vimalajeewa,sasi.bala}@waltoninstitute.ie
https://waltoninstitute.ie/

Abstract. Consumption of food in excess of the required optimal nutritional requirements has already resulted in a global crisis and this is from the perspective of human health, such as obesity, as well as food waste and sustainability. In order to minimize the impact of these issues, there is a need to develop novel innovative and effective solutions that can optimally match the food consumption to the demand. This requires accurate understanding of the food digestion dynamics and its impact on each individual's physiological characteristics. This study proposes a model to characterize digestive system dynamics by using concepts from the field of Molecular Communications (MC), and this includes integrating advection-diffusion and reaction mechanisms and its role in characterizing the digestion process as a communication system. The model is then used to explore starch digestion dynamics by using communication system metrics such as delay and path loss. Our simulations found that the long gastric emptying time increases the delay in starch digestion and in turn the glucose production and absorption into the blood stream. At the same time, the enzyme activity on the hydrolyzed starch directly impacts the path loss, as higher reaction rates and lower half saturation concentration of starch results in lower path loss. Our work can lead to provide insights formulated for each individuals by creating a digital twin digestion model.

Keywords: Digestion dynamics · Molecular communication · Advection-diffusion-reaction model

1 Introduction

The United Nations (UN) aims to halve the per capita global food waste at the retail and consumer levels by 2030 as ~1/3 of food produced globally for human consumption is wasted [4]. Metabolic food waste and food consumption in excess of the optimal nutritional requirements, uses valuable agricultural resources and results in critical health issues such as obesity. Besides, obesity in humans is also

---

Supported by *VistaMilk* Research Center, Ireland.

T. Nakano (Ed.): BICT 2021, LNICST 403, pp. 117–133, 2021.
https://doi.org/10.1007/978-3-030-92163-7_10

associated with ~20% greenhouse gas emissions that is relative to the normal-weight state estimated from increased food intake, aerobic metabolism and fossil fuel use for transportation [11]. These facts emphasis the need for innovative solutions to overcome such challenges and achieve the goals set by the UN.

Facilitating to set up more sustainable food consumption habits is one innovative solutions that is being widely investigated, where the optimal proportion of nutrition can be allocated to each individual based on their health condition as well as physiological setup and lifestyle. This facilitates consumers to design food that suits their personal nutritional needs, by considering factors such as age, gender, and physical fitness. Using this practice will help improve global human health and reduce food waste in a number of ways. For instance, setting up proper dietary plan and exercise routines to maintain optimal body weight. From a digital modeling perspective, digestive system dynamics can reveal valuable information which can be used to decide suitable food based on suitability to individual personal traits and health benefits [7]. Exploring digestive system dynamics can thus contribute to designing frameworks for setting up more sustainable food consumption habits, which will result in a more population-oriented food production and supply chain in the near future.

With the focus on expanding knowledge about the digestive system dynamics, this study aims to looks at its functionality from a different angle compared to the existing approaches, by proposing a mathematical model that can be used to a Molecular Communication (MC) system representation. A majority of previous studies has considered the use of computational models to characterize digestive system dynamics by using its physio-chemical properties. For instance, the study [9] provides different modeling methodologies, particularly compartmental models to characterize gastrointestinal tract dynamics during digestion. The study [8] also presents a compartment modeling approach using chemical kinetics in the context of drug administration, while authors in [12] proposes a set of computational models considering different properties of food which include completely soluble, non-soluble and non-degradable types. In addition, physiologically based Pharmacokinetics models have also been used to explore the effects of different physiological parameters such as age and disease status and then use derived insights for drug discovery and development [20]. This study takes a different direction by developing a computational model that uses concepts from MC, and used to mapped its functionality from a communication theory perspective. The recent introduction of MC has provided a new mechanism of characterizing biological cellular communication, and also facilitates new approaches of using engineered bionanoscience to control the communication process. From both pharmaceutics and food sciences perspectives, this can lead to development of innovative solutions for various critical issues such as targeted drug delivery for various nutritional diseases [3] and controlled delivery of nutrients [5].

In the model building process, three main stages of the digestive process are modeled as a MC system as illustrated in Fig. 1. First, mechanically broken down in the mouth goes to the stomach and then mixed with gastric juice stomach (*digesta*). The stomach releases digesta into the small intestine (this study terms this as the *small intestinal (SI) tract*). In the second stage, the digesta reacts with different enzymes available in the SI tract and converts it into an absorbable form (nutrients) while traveling along the SI tract. Finally, in the third stage, the SI wall selectively absorbs them into the circulatory system. These three stages are then mapped to an advection-diffusion, and reaction based MC system [13] (further explanation will be provided in the following section). The model is derived by considering the digestion of starch process. Exploring starch digestion dynamics is very important as it is the main source of carbohydrates in our diet and plays a crucial role as a source of energy. Carbohydrates supports fat metabolism as well as effective use of dietary proteins [16]. However, excess consumption of carbohydrates along with dietary fat can contribute towards critical health issues such as obesity. Thus, broader knowledge about carbohydrate digestion dynamics can help in diabetes and obesity management [19]. This study takes into account the starch digestion and absorption of glucose into the blood as a advection-diffusion and reaction based MC system. This is followed by understanding the impact of different parameters such as half-gastric emptying time (i.e., time to empty half of digesta volume) on the digestive system dynamics. Finally, performance metrics used in communication theory such as delay and path loss, are used to evaluate the performances of the starch digestion system.

The reminder of the study is organized as follows. Next Sect. 2 presents the system model and this is followed by the results derived through simulation studies using Python software simulation in Sect. 3. The importance of the results and their implications along with further studies are discussed in Sect. 4 and finally, Sect. 5 concludes the paper.

2   System Model

This section first gives a brief overview about the food digestion process and its mapping to a MC system. Next, a transport model for nutrient molecules in the SI tract is derived and then it is used to formulate the transformation of starch into glucose. Finally, two performance metrics delay and path loss are listed to evaluate model performance.

2.1   Food Digestion Process as a Communication System

In the digestive system, consumed food undergoes several intermediate stages before nutrients contained in food are absorbed into the circulatory system as illustrated in Fig. 1. The circulatory system then transports the nutrients and delivers to different parts of the body. The food digestion process is mapped to the components into a MC system as follows:

– **Stomach:** is the organ where food is mixed with gastric juice broken down mechanically and chemically before entering into the small intestine. Therefore, the stomach acts as a reservoir of gastric content.

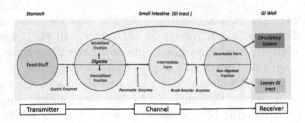

Fig. 1. An overview of the digestion process as a molecular communication system

- **Small Intestine (SI tract):** is the organ where most of the chemical digestion takes place due to enzymatic reactions. Its muscles' contraction and relaxation creates convective movement of the nutrient along the SI tract, while
  their absorption through diffusion. Its wall consists of *villi* that increases the
  surface area to absorb nutrients effectively.
- **Digestion:** is the process by which consumed food converts into absorbable
  nutrients through enzymatic hydrolysis. Different types of enzymes such as
  amylases, lipases and proteases help to hydrolyze dietary nutrients such as
  starch, lipid and protein. In particular, human diet contains a greater portion
  of carbohydrates of which starch is the main source. Starch is hydrolized into
  maltose, maltotriose and limit dextrins within the SI tract by pancreatic amy-
  lase, while some residual (resistant starch) pass into the large intestine where
  intestinal bacteria-based digestion takes place. The brush border enzymes
  (maltase-glucoamylase and sucrase-isomaltase) in the SI tract further break
  down these intermediate digestion products into glucose.

The digestive process described above are characterized and mapped into an
advection-diffusion and reaction based MC system as follows (see Fig. 1):

(a) **Transmitter:** Stomach acts as a transmitter as it releases digesta into the
    SI tract.
(b) **Channel:** Since the SI tract facilitates digesta to moving along the GI wall
    providing a medium to convert food particles into absorbable nutrients, the
    SI transport mechanism is represented as a MC channel.
(c) **Receiver:** The SI wall represents the receiver of a MC system as it selec-
    tively allows nutrient molecules to penetrate through the SI wall into the
    circulatory system.

2.2 Feed Stuff Transport Model

Figure 2 illustrates food particles (in terms of mass) entering the SI tract from
the stomach and the rate of mass flow or mass flux $Q(gm^{-2}s^{-1})$ through a
cross sectional area $A(m^2)$ in the SI tract. We also consider a control volume
of depth δx as $V = A\delta x$. The rate of change of the nutrient mass per unit
volume (concentration) $C(x,t)$ (gm^{-3}) through the control volume V (m^3) can
be expressed through the following mass balance expression.

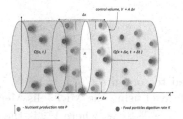

Fig. 2. Transport and production of nutrients during the small intestinal digestion of food particles.

$$V\frac{\partial C(x,t)}{\partial t} = \text{Mass}_{in} - \text{Mass}_{out},\tag{1}$$

where the terms, inlet and outlet mass of nutrients (Mass_{in} and Mass_{out}) in Eq. 1 are formulated as

$$\text{Mass}_{in} = Q(x,t)_{in}A + \text{Nutrient Production},$$
$$= Q(x,t)_{in}A + C_P(x,t)V,\tag{2}$$
$$\text{Mass}_{out} = Q(x+\delta x,t)_{out}A + \text{Nutrient Absorption},$$
$$= Q(x+\delta,t)_{out}A + K_aC(x,t)M,\tag{3}$$

where $C(x,t)_P$ $(gm^{-3}s^{-1})$ and K_a (ms^{-1}) are respectively the nutrient production and absorption rates, $M = (2\pi d\delta)f$ is the surface area of the controlled volume, where d is the radius and f is the surface area increased due to folds (villi) in the SI tract wall. By substituting Eq. 2 and 3 in 1, the change in nutrient mass can be expressed as

$$\frac{\partial C(x,t)}{\partial t} = -\frac{\partial Q(x,t)}{\partial x} - KC(x,t) + C_P(x,t),\tag{4}$$

where $K = \frac{2}{d}K_af$.

Considering the advection-diffusion based flow of nutrients in the SI tract, the mass flow rate of nutrients can be expressed as in Eq. 5 by using the Fick's first law.

$$Q(x,t) = C(x,t)u - D\frac{\partial C(x,t)}{\partial x},\tag{5}$$

where u (ms^{-1}) is the average velocity of the nutrient mass flow. Then, Eq. 6 expresses the partial derivative of Eq. 5 with respect to traveling distance, x, and is represented as

$$\frac{\partial Q(x,t)}{\partial x} = u\frac{\partial C(x,t)}{\partial x} - D\frac{\partial^2 C(x,t)}{\partial x^2},\tag{6}$$

Combining Eq. 4 and 6, the change in nutrient mass can be expressed as in Eq. 7. In general, this is known as the governing equation of the advection-diffusion and reaction based fluid flow in a pipe, and is represented as

$$\frac{\partial C(x,t)}{\partial t} + u(x,t)\frac{\partial C(x,t)}{\partial x} = D\frac{\partial^2 C(x,t)}{\partial x^2} - KC(x,t) + C_P(x,t). \tag{7}$$

Models from [15] are used to compute $C_P(x,t)$ and K_a. The nutrient production rate ($C_P(x,t)$) is derived using the Michaelis-Menten-kinetics and this is based on $C_P(x,t) = V_{max}\frac{C(x,t)}{K_{max}+C(x,t)}$, where V_m is the maximum reaction rate and K_{max} is the half saturation concentration. The nutrient absorption rate $K_a = 1.62\left(\frac{uD^2}{Ld}\right)^{\frac{1}{3}}$, $D = \frac{K_B T}{6\pi\mu r_m}$, where, D is the diffusion coefficient, K_B is the Boltzmann constant, T represents the absolute temperature and r_m is the radius of nutrient molecules and L and d are the length and diameter of the SI tract, respectively.

2.3   Starch Digestion Model

Following Eq. 7, starch digestion into glucose and absorption into blood in the SI tract are formulated below as a system of differential Eqs. 8–10. In this model, it is assumed here that the starch digestion starts after entering the SI tract. Also, the model is created using carbohydrate as the only source of nutrients, but humans and animals consume mixed macro-nutrients within diverse food matrices, which impact on the how the carbohydrates are digested. Thus, the presence of other macro-nutrients in the carbohydrate enriched diet significantly impacts on the glucose availability in the blood.

In response to the starch in the stomach $C_{st}(t)$ (sometimes the calls gastric volume) emptying at a rate γC_{st} (see Eq. 8, where $\gamma = \log(2)/t_{1/2}$ and $t_{1/2}$ is the half-gastric emptying time), will result in the starch concentration in the SI tract $C_s(x,t)$ increasing at the same rate. While starch molecules travel along the SI tract under the advection-diffusion mechanism, they react with the enzymes and consequently produces starch hydrolysis products (e.g., maltose) and then brush border enzymes (e.g., maltases) convert them into glucose. This study does not take into account concentrations of intermediate products and considers only the conversion of starch into glucose. Hence, the $C_s(x,t)$ decreases along the SI tract at a rate $V_m\frac{C_s}{K_m+C_s}$ (see Eq. 9). The glucose in the SI tract, $C_g(x,t)$, then increases at the same rate as the starch digestion. At the same time, it decreases at a rate $KC_g(x,t)$ due to the absorption of glucose molecules into the circulatory system while they travel along the SI tract under the same mechanism as the starch molecules (see Eq. 10). Please see [6] for more information about the digestive system modeling. Altogether, given that the boundary conditions $C_{st}(t = 0) = C_0(g)$, $C_s(x = 0, t = 0) = 0$ and $C_g(x = 0, t = 0) = 0$, the system of differential equations to characterize the starch digestion and production of glucose is expressed as

$$\frac{dC_{st}(t)}{dt} = -\gamma C_{st}(t), \tag{8}$$

$$\frac{\partial C_s(x,t)}{\partial t} = D_s \frac{\partial^2 C_s(x,t)}{\partial x^2} - u \frac{\partial C_s(x,t)}{\partial x} - V_m \frac{C_s(x,t)}{K_m + C_s(x,t)} + \gamma C_{st}(t), \tag{9}$$

$$\frac{\partial C_g(x,t)}{\partial t} = D_g \frac{\partial^2 C_g(x,t)}{\partial x^2} - u \frac{\partial C_g(x,t)}{\partial x} + V_m \frac{C_s(x,t)}{K_m + C_s(x,t)} - K C_g(x,t), \tag{10}$$

where $C_{st}(t)$ and $C_s(x,t)$ are respectively the starch concentrations in the stomach and SI tract and $C_g(x,t)$ is the glucose concentration in the SI tract and D_s and D_g are the diffusion coefficients of starch and glucose, respectively.

2.4 Analytical Solution

Based on the differential Eqs. 8–10, we derive analytical solutions assuming $K_m << C_s(x,t)$. In reality, this assumption emphasizes that nearly all enzymes will be occupied by the starch molecules. Hence, this assumption leads to a simplification that $V_m \frac{C_s}{K_m + C_s} \approx V_m$. The analytical solution of three Eqs. 8–10 are given as follows

(1) **Starch in Stomach:** Suppose at $t = 0$, C_0 amount of starch is consumed, then the solution of Eq. 8 represents the variability in starch concentration in the stomach, C_{st} over time, and is represented as

$$C_{st}(t) = C_0 e^{-\gamma t}. \tag{11}$$

(2) **Starch in SI tract:** By substituting $C_s(x,t) = f(x,t)e^{\alpha x - \beta t} + pe^{\gamma t} + qV_m t$ (alpha, β, p and q are parameters to be evaluated) in Eq. 9, its solution can be represented as

$$C_s(x,t) = \frac{f_s}{\sqrt{4\pi Dt}} e^{\frac{-(x-ut)^2}{4D_s t}} - C_0 e^{-\gamma t} - V_m t. \tag{12}$$

where $f(x,t) = [C_s(x,t) + C_0 e^{-\gamma t} + V_m t] e^{\left(\frac{u^2}{4D_s}t - \frac{u}{2D_s}x\right)}$ and $f_s = f(0,0)$.

(3) **Glucose in SI tract:** Following the similar approach as in the step 2 with the substitution $C_s(x,t) = f(x,t)e^{\alpha x - \beta t} + pV_m t$ (alpha, β, and p are parameters to be evaluated), the solution of Eq. 10 can be expressed as follows.

$$C_g(x,t) = \frac{f_g}{\sqrt{4\pi Dt}} e^{\frac{-(x-ut)^2}{4D_g t} - Kt} + V_m t, \tag{13}$$

where $f(x,t) = [C_g(x,t) - V_m t] e^{\left[\left(\frac{u^2}{4D_g}+K\right)t - \frac{u}{2D_g}x\right]}$ and $f_g = f(0,0)$.

Table 1. Model parameters used for simulations.

Parameter	Symbol	Value
Small intestine length	l	6.9 m (male)
Small intestine diameter	d	2.5 cm
Viscosity	μ	0.01–10 Pas
Radius of Starch molecules	$r_m s$	5 nm
Radius of Glucose molecules	$r_m g$	0.38 nm
Surface area increase due to fold, vili and microvili	f	12
Mean velocity	u	1.7×10^{-4} ms^{-1}
Maximum reaction rate	V_{max}	25 mM min^{-1}
Half saturation concentration	K_{max}	9 mM
Half gastric emptying time	$t_{1/2}$	1 h

2.5   Channel Characteristics

Two performance metrics listed below are used to evaluate the molecular communication system of the starch digestion dynamics in the SI tract.

(1) **Path Loss (PL):** This is considered as the loss of starch or glucose traveling from an arbitrary distance x_1 to x_2 ($0 \leq x_1 < x_2 \leq l$) along the SI tract. By following the path loss equation given in [14], PL is computed as

$$PL = 10 \log_{10} \left(\frac{C_g(x_1, t)}{C_g(x_2, t)} \right). \tag{14}$$

(2) **Delay:** It is generally assumed that a significant amount (>90%) of consumed carbohydrates enters into the circulatory system as glucose within 2–3 h. In this study, we assumed that 1.5 h as the benchmark to digest >50% of consumed carbohydrates, the delay is computed as the difference between the time taken to digest and then absorb consumed carbohydrates and the benchmark time.

3   Results

The model derived in the previous section is used here to characterize starch digestion dynamics along the SI tract through simulations studies which were performed by using the model parameters listed in Table 1 and they were taken from [15] and [17]. The starch molecule size is used here is for rice starch.

3.1   Starch Digestion and Production of Glucose

As mentioned in the previous section, starch is converted into glucose while traveling along the SI tract so that both the starch and glucose concentrations

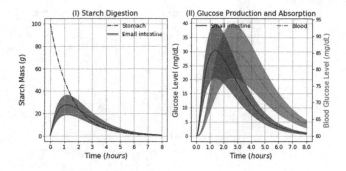

Fig. 3. Having consumed 100 g of carbohydrates, change in starch and glucose concentrations in the stomach as SI tract.

vary over the SI tract with respect to the distance that they travel over time. Figure 3 shows the variability in the starch concentrations averaged over distance in both the stomach and SI tract along with the glucose concentration, given that 100 g of carbohydrates is consumed and the normal blood glucose level is 60 mg/dL. As Fig. 3(I) depicts, starch mass in the stomach decreases over time and a greater fraction of consumed carbohydrates has been entered into the SI tract within 1–1.5 h. As a consequence, initially, the starch concentration in the SI tract depicts an increase and then it gradually decreases in response to the degradation of starch into glucose. As a result, in Fig. 3(II), the glucose concentration in the SI tract increases and the maximum increase is around 30 mg/dL. In response to the absorption of glucose into blood, the blood glucose level increase and then decreases as glucose is stored in the body. The Beragman's minimal model [?] was used to compute the blood glucose level.

In general, Fig. 3 shows a greater portion (>90%) of starch has been digested and the produced glucose has been absorbed into the blood around in 2–3 h and the observed maximum glucose change is around 30 mg/dL.

3.2 Effectiveness of Starch Digestion

The effectiveness of the starch digestion process could, however, be affected by a number of factors. This results in delay in production and absorption of glucose which is used as the main source of energy required for different body functions. Figure 4 shows the impact of selected parameters (consumed quantity, stomach emptying time, viscosity and velocity) on the starch digestion process. According to Fig. 4a, the time taken to absorb the produced glucose into blood is slightly increasing with the increase in consumed quantity of carbohydrates. On the other hand, Fig. 4b depicts that the time taken to digest 100 g of carbohydrates and then absorbing the produced glucose into blood is higher with longer stomach emptying time. In addition, Fig. 4c shows the change in glucose concentration in the SI tract in response to the increase in velocity (u) and viscosity (μ) of glucose molecules in the SI tract, given that 100 g of carbohydrates has been consumed. In the model, the increase in viscosity results in increasing glucose

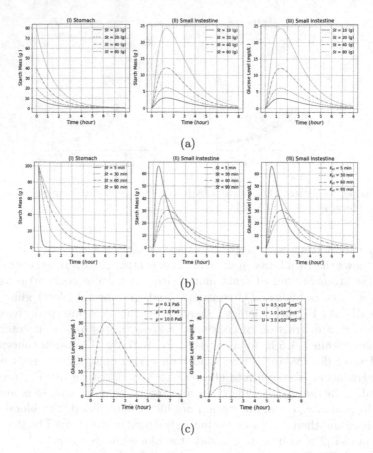

Fig. 4. Variability in starch and glucose concentration in the stomach and SI tract with respect to the (a) amount of consumed carbohydrates, (b) and the half stomach emptying time, and (c) change in viscosity (μ) and velocity (u) (only in the SI tract).

level as higher viscosity slows down the starch movement speed and this in turn, allows starch molecules to spend long period of time in the SI tract. Conversely, increasing velocity decreases the time spent by starch molecules in the SI and hence, results in decrease of the glucose level.

3.3   Delay in Starch Digestion

For a healthy person, it is generally known that carbohydrates digestion results in a sudden increase in blood glucose level and on average takes an hour to reach its peak and then it stabilizes in response to the secretion of insulin from the pancreas. Figure 4 showed that different factors can have a significant influence on delaying the digestion process. The stomach emptying time acts as one of the most influential factors among them because long stomach emptying time can result in significant delay in the remaining digestion process. So, this study

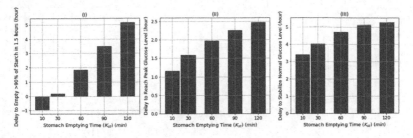

Fig. 5. The impact of stomach emptying time on the starch digestion delay and corresponding delay it creates in stabilizing blood glucose level.

explored only the impact of the stomach emptying time and Fig. 5 displays the impact of this on delaying the starch digestion process. Given the average time taken to digest >50% of consumed carbohydrates is 1.5 h, Fig. 5(I) shows the delay of starch entering from the stomach into the SI tract with increasing stomach emptying time. The corresponding increase in time to reach maximum blood glucose level and reaching back to normal level (stabilizing glucose level) is displayed respectively in Fig. 5(II) and 5(III). Therefore, as seen in Fig. 5, longer stomach emptying time delays the starch digestion process, which in turn, could cause delaying the body glucose demand required to produce energy. Consequently, such a delay may lead to certain disorders in different body functioning.

3.4 Path Loss with Respect to Starch and Glucose

The PL can result from both the loss in undigested starch and unabsorbed glucose in the small intestine. Figure 6 shows the variability in PL in response to the change in velocity u and half saturation concentration K_{max} over the range of $[0.01–10] \times 10^{-4}$ m/s and $[0–40]$ mM, respectively. As displayed in Fig. 6a, the PL with respect to starch does not show significant change with increasing velocity, but increasing K_{max} contributes to decreasing PL at a declining rate. PL with respect to glucose given in Fig. 6b shows an increase with increasing both u and K_{max} and also the increasing rate with u is greater compare to that with K_{max}, but the increment is declining with larger u. Besides, Fig. 7 shows the impact of V_{max} and K_{max} together on PL is depicted while V_{max} and K_{max} are varying over the ranges $[10–30]$ mM/min and $[0–40]$ mM, respectively. The PL with respect to both starch and glucose shows decreasing trend with respect to increase in V_{max}, but they show an increasing trend with increasing K_{max}.

4 Discussion

Exploring starch digestion dynamics is important for a number of reasons. It is the main source of carbohydrates in our diet and acts as the highly efficient energy source required for body function. Thus, efficient carbohydrate digestion is vital, for instance in intensive exercises as it avoids the body using protein

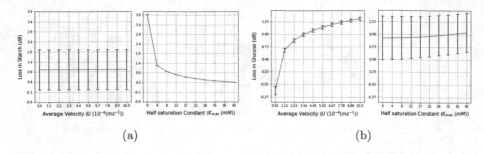

Fig. 6. Path loss with respect to starch and glucose while varying velocity u and K_{max} together.

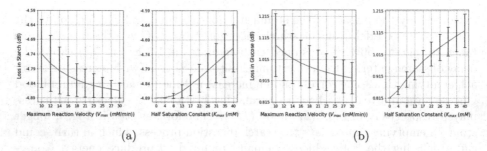

Fig. 7. Path loss with respect to starch and glucose while varying maximum reaction rate V_{max} and half saturation concentration K_{max} together.

stored in the internal organs to produce energy. This in turn helps preserving lean muscle mass and the use of dietary protein for other important purposes such as building muscles rather than consuming for generating energy. Moreover, carbohydrate helps the function of brain and nervous system and metabolism of fat. Effectiveness of starch digestion could, however, be varied from person to person due to variability in influenced factors such as gastric emptying times, flow of digestive content, digestive capacity and efficacy of glucose transporters, as well as the glucose absorption into the blood [1]. Thus, any abnormality of one or more of these factors could potentially lead to creating potential disorders in aforementioned various body functions interconnected with glucose. For instance, imbalance in optimum blood glucose level could cause a number of issues because high plasma glucose level could leads to microvascular and neural damages, and renal inefficiency while low glucose level causes anxiety, aggressiveness, coma and eventually death [18]. Therefore, efforts to understand influence of these factors would be therapeutically important.

The MC-based approach proposed can provide insights which can contribute to characterizing the influence of such factors on the starch digestion process effectively. It would be useful for taking actions to control diabetes and obesity, for example. Therefore, this section discusses the significance of these insights for the transmitter, channel and receiver through simulation studies that can contribute to design of appropriate interventions.

4.1 Transmitter: Effective Gastric Emptying Time

In the digestion system, effectiveness of functions associated with the transmitter (i.e., stomach) is a crucial entity that can control the whole starch digestion process. This is because the glucose absorption into blood depends on the rate at which gastric content is released into the small intestine. Since the gastric emptying rate γ is computed by using the time to empty the stomach content by half of its volume, it directly influences on the delay in the whole starch digestion process. Various factors influence the increasing gastric emptying rate and this can include high volume, viscosity, exercise level and relative nutrient contents in the meal. Moreover, aging also increases this value, which in turn, decreases the efficacy of absorption of glucose and also the calcium and iron too [10]. These dynamics can easily be characterized by using the MC-based approach as it allows exploring variability in starch and glucose concentrations in the small intestine with respect to change in certain model parameters. For instance, both Fig. 4 and 5 shows the potential of the proposed model in exhibiting the impact of variability in delay with increasing stomach emptying time, velocity and viscosity. Increasing delay in starch digestion could directly influence the homeostasis in blood glucose level which is crucial for optimum body functions. Therefore, these insights could be incorporated for exploring dynamics in insulin sensitivity and glucose effectiveness which are very important for controlling blood glucose level, as well as designing personalized treatments for diabetes patients.

4.2 Channel: Efficacy in Mechanical and Chemical Functions

Digestive dynamics within the small intestine can primarily be characterized by its mechanical and chemical functionalities. The mechanical functionalities include movements in muscles while enzymatic reactions include the chemical functionalities. Variability in these processes can thus impact effectiveness of the digestion process.

The contraction and relaxation of muscles in the SI tract help effective separation, mixing and propelling of digesta, which in turn, creates convective flows allowing effective transformation of digesta into absorbable nutrients. The Brownian motion of nutrient molecules creates diffusion-based movement which helps nutrient absorption into the blood. Optimal functionality in these processes allows sufficient transient time that is essential for complete digestion and increasing the nutrients absorbed into blood. In the model discussed here, bulk flow velocity (u) and viscosity (μ) are two key parameters which can make a greater influence on the movement of nutrient molecules. For this reason, Fig. 4c shows a greater glucose production with low velocity and high viscosity. Figure 7a also supports this observation, indicating a lower path loss with respect to smaller velocity.

With regards to the chemical reactions, the model considered only the enzymatic action on the starch molecules based on Michaelis-Menten kinetics. Due to the complexity in solving the differential equations, the analytical solutions were derived considering the half saturation concentration (K_{max}) is very small compared to the starch concentration. This means all enzymes will have been occupied by starch molecules by the saturation stage of the reaction between the starch and enzymes. Therefore, Fig. 6 shows opposite trends in loss with respect to the starch and glucose with increasing K_{max}. Moreover, Fig. 7 suggest that smaller half saturation concentration results in greater contribution to reduced path loss.

4.3   Receiver: Effective Nutrient Absorption

Lower path loss with respect to both starch and glucose can contribute to increasing the amount of nutrients detected by the receiver (i.e., the small intestine wall). According to the model, nutrient absorption also depends on bulk flow size of nutrient molecules, velocity, and viscosity as parameters related to small intestine such as length and diameter are fixed. Therefore, as discussed above, short gastric emptying time, low velocity and high viscosity can contribute to improving the glucose absorption effectiveness. The impact of nutrient molecule size was not considered since this study considered glucose absorption only. Thus, path loss has a potential to imply the effectiveness of the nutrient absorbance depending on variability in physiological factors of the SI tract with respect to age, gender, and health conditions. Therefore, such insights could be helpful, for instance, in making dietary plans depending on personal requirements.

4.4   Future Directions

Since the study lays a strong foundation to explore the digestive dynamics in terms of MC concepts, it unveils a number of ways that this study could be further extended. Most importantly, given the amount of glucose absorbed into the blood, variability in blood glucose level in the body can be explored with respect to a range of factors such as age, gender, and physical fitness. This can help in terms of designing therapeutic treatments for chronic diseases such as diabetes. Moreover, it would also help in taking actions against obesity as controlling carbohydrates intake that can effectively stimulate burning excess body fat by increasing activity levels. In addition, males and females differ in their food and calorie intake, with males generally consuming more food including carbohydrates than females. The model discussed here is general and not specific to any gender, but there is a potential of using it to explore gender-based variability in digestive dynamics by including the gender related factors into the model parameters, such as the gut length and need of calories.

This study accounts the impact of few parameters on the starch digestion process so that it is able to provide an overall idea about the digestion dynamics. This is because the enzymatic activity depends on the pH value and the temperature but the study assumes they do not vary over the digestive tract though they have a significant impact on the digestion process. The model assumes the gut width is fixed but it varies along the tract, and by the development stage as the individual grows, so does the size of the gut including the width and length, which can accommodate increased intake to support growth.

Also the starch digestion starts in the mouth. However, this study assumed the starch stays intact until it reaches the small intestine. The model was created using carbohydrate as the only source of nutrients, but humans and animals consume mixed macronutrients within diverse food matrices, which impact on the how the carbohydrates are digested. The presence of other macronutrients in the carbohydrate enriched diet significantly thus impacts on the glucose availability in the blood. Taking into account these factors in the model development process will also contribute to improving the potential of creating an accurate digestive dynamics digital model.

5   Conclusion

This study looks at the digestive system functionality as a MC system and then proposes an advection-diffusion and reaction based model to characterize digestion system dynamics in the context of starch digestion. Based on communication system evaluation, two performance metrics, which are delay and path loss are used to explore the influence of different properties of the digestive system in the context of starch digestion. According to the observations derived from simulation of an analytical model, the gastric emptying rate, velocity and viscosity of digesta along with the parameters related to enzymatic activity on the starch such as half saturation concentration influences on the delay and path loss in the starch digestion process. These observations primarily suggest that shorter gastric emptying time with low velocity and high viscosity of digesta can contribute to improving the amount of glucose absorbed into the blood. Therefore, these insights help in expanding the knowledge about digestive system dynamics so that they can contribute towards the development of novel solutions for future food-related global crisis such as obesity and food waste. Most importantly, the study lays a concrete foundation to drive digestive system studies towards a new direction that will enable us to create a digital twin that will personalize our ingredients that suits our personal physiological settings as well as internal digestion functions.

Acknowledgment. This research was supported by Science Foundation Ireland and the Department of Agriculture, Food and Marine on behalf of the Government of Ireland VistaMilk research centre under the grant 16/RC/3835.

References

1. Anthony, C.D., Guilhem, P., Robert, G.G., Philip, W.K.: Digestion of starch: in vivo and in vitro kinetic models used to characterise oligosaccharide or glucose release. Carbohydr. Polym. **80**(3), 599–617 (2010). https://doi.org/10.1016/j.carbpol.2010.01.002

2. Bergaman, R.N., Ider, Y.Z., Bowden, C.R., Cobell, C.: Quantitative estimation of insulin sensitivity. Am. J. Physiol. **236**(6), 667–677 (1979). https://doi.org/10.1152/ajpendo.1979.236.6.E667
3. Chahibi, Y.: Molecular communication for drug delivery systems: a survey. Nano Commun. Networks **11**, 90–102 (2017) https://doi.org/10.1016/j.nancom.2017.01.003, https://www.sciencedirect.com/science/article/pii/S1878778917300054
4. Christel, C., Ulf, S., van Otterdijk, R., Alexandre, M.: Global food losses and waste. Food and Agriculture Organization of the United Nations (FAO) (2011). http://www.fao.org/3/mb060e/mb060e.pdf
5. Dixon, V., Subhasis, T., Breslin, J., Donagh P., B., Sasitharan, B.: Block chain and internet of nano-things for optimizing chemical sensing in smart farming (2020)
6. Gregersen, H., Kassab, G.: Biomechanics of the gastrointestinal tract. Neurogastroenterol Motil. **8**, 277–297 (1996). https://doi.org/10.1111/j.1365-2982.1996.tb00267.x
7. Hannelore, D., Pamela, B., Monique, R.: Smart personalized nutrition: Quo vadis. In: Smart Personalized Nutrition Workshop. European Commission FOOD 2030 (2012)
8. Khanday, M., Rafiq, A., Nazir, K.: Mathematical models for drug diffusion through the compartments of blood and tissue medium. Alexandria J. Med. **53**(3), 245–249 (2017). https://doi.org/10.1016/j.ajme.2016.03.005
9. Le Feunteun, S., Al-Razaz, A., Dekker, M., George, E., Laroche, B., van Aken, G.: Physiologically based modeling of food digestion and intestinal microbiota: state of the art and future challenges. an infogest review. Ann. Rev. Food Sci. Technol. **12**(1), 149–167 (2021). https://doi.org/10.1146/annurev-food-070620-124140. pMID: 33400557
10. Leiper, J.B.: Fate of ingested fluids: factors affecting gastric emptying and intestinal absorption of beverages in humans. Nutr. Rev. **73**, 57–72 (2015). https://doi.org/10.1093/nutrit/nuv032
11. Magkos, F., et al.: The environmental foodprint of obesity. Obesity **28**(1), 73–79 (2020). https://doi.org/10.1002/oby.22657
12. Masoomeh, T., Philippe, L., Jean-René, L., Christine, G., Guy, B.: Mathematical modeling of transport and degradation of feedstuffs in the small intestine. Theor. Biol. **294**, 114–121 (2012). https://doi.org/10.1016/j.jtbi.2011.10.024
13. McGuiness, D.T., Giannoukos, S., Marshall, A., Taylor, S.: Experimental results on the open-air transmission of macro-molecular communication using membrane inlet mass spectrometry. IEEE Commun. Lett. **22**(12), 2567–2570 (2018). https://doi.org/10.1109/LCOMM.2018.2875445
14. Mohammad Upal, M., Dimitrios, M., Mouftah, H.: Characterization of molecular communication channel for nanoscale networks. In: 3rd International Conference on Bio-inspired Systems and Signal Processing (BIOSIGNALS-2010), pp. 327–332 (2010)
15. Moxon, T., Gouseti, O., Bakalis, S.: In silico modelling of mass transfer and absorption in the human gut. J. Food Eng. **176**, 110–120 (2016). https://doi.org/10.1016/j.jfoodeng.2015.10.019
16. Pesta, D.H., Varman, T.S.: A high-protein diet for reducing body fat: mechanisms and possible caveats. Nutr. Metab. **11**, 1–53 (2014). https://doi.org/10.1186/1743-7075-11-53
17. Singh, J., Kaur, L.: Chapter 8 - potato starch and its modification. In: Advances in Potato Chemistry and Technology (Second Edition), 2nd edn. pp. 195–247. Academic Press, San Diego (2016). https://doi.org/10.1016/B978-0-12-800002-1.00008-X

18. Sue, P., Lauretta, Q., Mary, B., Carol, F., Michael, M., Poul, S.: OES glycemic variability impact mood and quality of life? Diab. Technol. Ther. **14**(4), 303–310 (2012). https://doi.org/10.1089/dia.2011.0191
19. Zhang, B., Li, H., Wang, S., Junejo, S.A., Liu, X., Huang, Q.: In vitro starch digestion: mechanisms and kinetic models. In: Wang, S. (ed.) Starch Structure, Functionality and Application in Foods, pp. 151–167. Springer, Singapore (2020). https://doi.org/10.1007/978-981-15-0622-2_9
20. Zhuang, X., Lu, C.: Pbpk modeling and simulation in drug research and development. Acta Pharmaceutica Sinica B **6**(5), 430–440 (2016). https://doi.org/10.1016/j.apsb.2016.04.004, https://www.sciencedirect.com/science/article/pii/S221138351630082X. Drug Metabolism and Pharmacokinetics

Smart Tumor Homing
for Manhattan-Like Capillary Network
Regulated Tumor Microenvironment

Yin Qing[1] , Yue Sun[1,2](✉) , Yue Xiao[1] , and Yifan Chen[2]

[1] Chengdu University of Technology, Chengdu 610059, China
[2] University of Electronic Science and Technology of China, Chengdu 611731, China

Abstract. This paper investigates the tumor microenvironment regulated by the dense interconnected capillary network nearby, forming Manhattan-like biological gradient fields (BGFs) distribution. The research to date has tended to focus on modeling in Euclidean space rather than Manhattan space. Based on the Manhattan-like BGFs, we propose a coordinate gradient descent (CGD) iterative algorithm to realize tumor homing. Nanorobots with contrast agents and sensors are employed as computing agents. The sensors serve as local sensing agents to provide iterative information, and the contrast agents can deposit themselves on the tumor through the enhanced permeability and retention (EPR) to make magnetic resonance imaging (MRI) easier to detect. We aim to achieve tumor homing using as few iterations as possible in a Manhattan-like BGF. The simulation results show that the proposed CGD algorithm has higher efficiency and fewer iterations in Manhattan-like BGFs than the brute-fore.

Keywords: Tumor homing · Manhattan-like BGFs · CGD iterative algorithm

1 Introduction

Magnetic Nanoparticle-mediated drug delivery has drawn much attention in cancer theranostics. Since the presence of magnetic nanoparticles (MNPs) introduces inhomogeneities in the local magnetic field that increase the signal decay rate, resulting in a significant contrast change in the tissue enriched with MNPs [1]. Thus, magnetic resonance imaging (MRI) employs MNPs as contrast agents to enhance tumor detection [2]. The main problem facing this approach is the targeted delivery relies on systemic blood circulation and only less than 2% of the nanoparticles can deliver to a precise site [3]. Another alternative approach is an iterative-optimization-inspired direct targeting strategy (DTS) for nanosystems [4]. Nanoswimmers assembled by many magnetic nanoparticles are served as computing agents. In the presence of an external magnetic field, the

T. Nakano (Ed.): BICT 2021, LNICST 403, pp. 134–144, 2021.
https://doi.org/10.1007/978-3-030-92163-7_11

nanoswimmers are guided to potential sites of tumors. When the tumor binds to the contrast agents, the enhanced permeability and retention (EPR) effect will be triggered, thus enabling tumors to be detected by MRI [5]. One of the other methods is a self-regulated and coordinated targeting strategy, inspired by the collaborative movements of the living cells in different natural biological processes [6]. This method doesn't require an external centralized control unit, nanoparticle swarm cooperates and coordinates autonomously through the information provided by the biological gradient fields (BGFs), and carries out tumor homing in the search domain.

The growth of tumors affects the microenvironment around, changing its physical properties. For instance, 60% of solid cancers contain hypoxic areas heterogeneously dispersed throughout the tumor [7]. Some tumors rely on glycolysis to produce energy, which leads to hypoxic concentration and acidosis, increasing the spatial distribution differences in tumor areas [8]. The biophysical information, regulated by capillary networks, is presented as a multi-dimensional BGFs distribution. The two alternative methods mentioned above both use BGFs as objective functions to provide iterative direction information and consider it as a continuous distribution in a Euclidean space.

This paper introduces BGFs similar to taxicab geometry. To deliver the oxygen to the tissues efficiently, capillaries shape as an even-distributed, two-dimensional grid network, as shown in Fig. 2(a) [9]. The formation of BGFs is related to the capillary network. The oxygen content levels in the blood decrease as the distance from the arterial end increases [10]. Therefore, tumor cells located at the distal end of the blood vessel may be hypoxic, even if they are close to the vessel [11]. Due to the lack of oxygen, cells produce lactic acid anaerobic respiration, resulting in a decrease in PH [12]. The formation of these BGFs has been inseparable from the capillary network, which is a Manhattan space. Therefore, so when considering an actual BGF, a Manhattan space should more suitable than Euclidean space.

Recent work improves the detection probability and targeting efficiency employing external magnetic field control of nanorobots [4]. The major advantage of tumor sensitization through external manipulation is that the route to reach the lesion tissues is optimal and minimization the risk of exposure to the systemic circulation. A similar and holistic approach is employed, integrating nanorobots with sensors, an external centralized control unit, and Manhattan-like BGFs for a tumor homing. We modify objective functions into Manhattan-like objective functions, the coordinate gradient descent (CGD) is the iterative search algorithm in a Manhattan space. For the acquisition of the iterative direction, using sensors to local sensing in vivo, an external controlling and tracking system, such as an integrated device consisting of pairs of electromagnetic coils to generate a rotating magnetic field [13] to drive the nanorobots.

The paper is organized as follows. In Sect. 2, potential errors of BGFs being an objective function are discussed, and the transformation from Euclidean metric objective functions to Manhattan-like objective functions are presented. Section 3 demonstrates the CGD algorithm and the acquisition of discrete gradient. Next, numerical simulations are performed to validate the effectiveness of the proposed algorithm in Sect. 4. Finally, Sect. 5 consists of the conclusion of this paper and future works.

2 Manhattan-Like Objective Function

BGFs can be divided into two types, environment-response and environment-primed. The first type can be obtained by using the physical properties of the tumor microenvironment, such as low pH, alteration in oxygen content, and blood velocity [10]. Special nanoparticles are injected into the blood flow to create an artificial BGF accumulated in the vicinity of the tumor. For instance, modified gold nanorods can accumulate in tumors through systemic circulation passively [14].

Tumor-induced BGFs can be used as auxiliary information for in vivo computing and provide evaluation criteria for position information of nanorobots. To understand and measure BGFs computationally, the objective functions with global maximum are considered to represent the possible situation of BGFs of in vivo computation. Assume that $\mathbb{D}$ denotes the high-risk area of the tumor. Then computing agents are represented by G. The objective functions based on Manhattan- and Euclidean-space are represented by F and f, respectively. As computational agents of in vivo computing, nanorobots are composed of natural materials that will physically, chemically, and biologically react with living tissues, which indicates that the interactions between agent G and the domain $\mathbb{D}$ induces disturbance, resulting in measurement error. Therefore, the objective function can be described as follows:

$$F(\boldsymbol{x}; G) = F_T(\boldsymbol{x}) + \epsilon(\boldsymbol{x}; G) \qquad \boldsymbol{x} \in \mathbb{D} \tag{1}$$

where $F_T(\boldsymbol{x})$ is the true objective function value at $\boldsymbol{x}$. $\epsilon(\boldsymbol{x}; G)$ is the random compensation error, which attempts to counteract the disturbance caused by G.

As mentioned above, BGFs in the capillary shape like a grid model, which further refers to a Manhattan space. Currently, the functions used in many standard testing problems are mostly Euclidean metrics. However, the test function in the Euclidean metric is continuous and derivable everywhere, unlike the Manhattan metric, which is defined only orthogonally. Thus, the Euclidean metric is not suitable for describing the proposed Manhattan-like BGFs. Therefore, we modify a Euclidean metric-based objective function into Manhattan-based. In order for our modifications to have more practical implications, the transformation derives from the physical definition differences between the Manhattan and Euclidean metric, which indicates that area and distance formed by BGFs are different in Manhattan space and Euclidean space.

Give two points (x_n, y_n), (x_m, y_m) and a expression as:

$$(x_m - x_n) * (y_m - y_n) \tag{2}$$

at the Euclidean metric-based (2) can represent the rectangular area with side lengths $\sqrt{(x_m - x_n)^2}$ and $\sqrt{(y_m - y_n)^2}$. Recalculated using Manhattan metric-based definition, the side length can be expressed as $|x_m - x_n|$ and $|y_m - y_n|$. Therefore, the area of rectangle can be modified as:

$$|x_m - x_n| * |y_m - y_n| \tag{3}$$

Similarly, the mapping table between Euclidean metric-based and Manhattan metric-based is given in Table 1 as a reference for modifying a Manhattan-like function.

Table 1. Mapping table from euclidean metric-based to manhattan metric-based.

Euclidean metric-based function	Manhattan-like function
$\sqrt{x^2 + y^2}$	$|x| + |y|$
x	$|x|$
$x - y$	$|x| - |y|$
$x * y$	$|x| * |y|$
x^3	$|x|^3$
x^2	$|x|^2$

Different objective functions are employed to describe the BGFs of the tumor in related works. In this paper, these objective functions in Euclidean space are modified to Manhattan-like functions, with the maximum and minimum value being normalized to 1 and 0, respectively. Then the selected search field is limited to a square with a side length of 10 mm.

1) Bowl-shaped

$$f(x, y) = 1 - 0.14 \left(\sqrt{x^2 + y^2} \right) \qquad (x, y) \in \mathbb{V} \tag{4}$$

The modifying result is as follows:

$$F(x, y) = 1 - 0.14(|x| + |y|) \qquad (x, y) \in \mathbb{V} \tag{5}$$

2) Plate-shaped

$$f(x, y) = 1 - \left(0.01 \left(x^2 + y^2 \right) + 0.02xy \right) \qquad (x, y) \in \mathbb{V} \tag{6}$$

The modifying result is as follows:

$$F(x, y) - 1 - \left(0.01 \left(|x|^2 + |y|^2 \right) + 0.02|x||y| \right) \qquad (x, y) \in \mathbb{V} \tag{7}$$

where $\mathbb{V}$ represents the capillary grid network. As shown in Fig. 1, the modified Manhattan-like functions are taken as the objective functions where the maximum value represents the location of the tumor. The numebr 0.14 reported in both (4) and (5), and the numbers 0.01 and 0.02 reported in (6) and (7) are the normalization coefficients, they do not affect the performance of the algorithm when the objective functions are used as test functions. The contour plots of the modified Manhattan-like functions are diamond-shaped, which is consistent with the situation in a Manhattan space. The gradient of the Bowl-shaped function is smooth and uniform, considered as the best case of BGFs. And the gradient of the Plate-shaped function is not as smooth as Bowl-shaped, therefore it can be treated as a general case of BGFs.

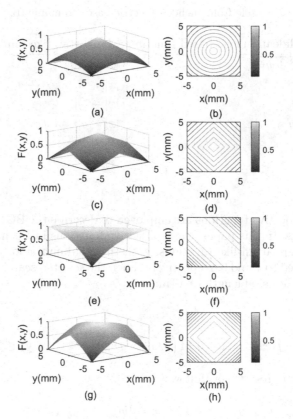

Fig. 1. Graphical representation of two objective functions. (a) Bowl-shaped and (b) its contour plot; (c) Bowl-shaped modified to manhattan-like results and (d) its contour plot; (e) Plate-shaped and (f) its contour plot; (g) Plate-shaped modified to manhattan-like results and (h) its contour plot. All objective functions have a maximum value of 1 and a minimum value of 0.

The capillary network consists of straight and rigid cylindrical blood vessels [15], and fractal models are often used to describe their bifurcation structure. However, fractal models cannot imitate the interconnected structure of the capillary network, especially in the vicinity of the tumor, which can be more appropriately described with a grid model [9]. In our model, the capillary is regarded as a grid model with an equal side length Δd. Then, the Manhattan-like objective function is mapped to the grid model. As shown in Fig. 2(b), only four vertices of each grid are used as the mapping results. The blue circles, represent possible path points after mapping, the red trajectory represents the iterative path of the agents.

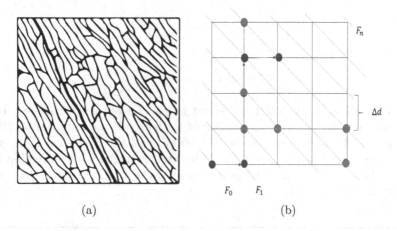

(a)   (b)

Fig. 2. (a) Blood vessels in microvasculature [9]; (b) Capillary network grid model and points on the grid that intersects $F_0 - F_n$ represent the equipotential points.

3   Coordinate Gradient Descent

The previous section has shown that the grid model of the capillary network is a Manhattan space. Therefore, it is necessary to adopt a Manhattan detection trajectory, which means subject the detect direction of the agents to limitations. Assuming that the blood flow direction is from lower left to upper right, then the actual detect direction of agents becomes upward or right. Thus, we propose the CGD iterative algorithm based on the gradient descent algorithm [4] and coordinate descent [16], which limits the direction of the gradient to the direction of the coordinate axis in order to conform to the detection of a Manhattan space.

Firstly, the discrete derivatives can be described as:

$$F'(x) = \frac{F(x + \Delta d) - F(x)}{\Delta d} \tag{8}$$

where Δd is the grid width as shown in Fig. 2(b). $F(x + \Delta d)$ is the value of objective function at location $x + \Delta d$, measured by the agents. The iteration position of agents is updated as follows:

$$x_{k+1} = x_k + \Delta x + s_{k+1} u_{\angle \varphi + \Delta \varphi} \tag{9}$$

where Δx is the location error, $u_{\angle \varphi}$ is a unit vector with a direction angle φ. In CGD the specific value of φ is 0 or $\frac{\pi}{2}$. $\Delta \varphi$ is the angular deviation, due to the steering magnetic field and agents steering imperfections. We simply assume that Δx, $\Delta \varphi$ are all subject to the variance of $\sigma_{\Delta x}^2$, $\sigma_{\Delta \varphi}^2$ and the mean of 0 normal distribution. s_{k+1} is the step size of the k+1-th iteration, the AdaGrad is employed to calculate the step size for each iteration. The step size updates as follows:

1) Accumulated derivative square

$$r_k = r_{k-1} + (F_k')^2 \tag{10}$$

2) Update step size

$$s_k = s_{k-1} - F_k' \left(\frac{s_{k-1}}{\delta + \sqrt{r_k}} \right) \tag{11}$$

where $r_0 = 0$, F_k' is the discrete derivative can be obtained from (8). In the calculation process, to avoid the value of r_k being 0, a constant $\delta = 10^{-7}$ is added to the denominator to ensure the stability of the numerical value, which will not affect the update of step size. The update in the y-axis direction is the same as that in the x-axis. The pseudocode of the CGD iterative algorithm is as shown in Algorithm 1.

Algorithm 1. Coordinate gradient descent (CGD)

Input: Objective function $F(x, y)$, initialize agent location: x_0, y_0, initial step size : s, constant $\delta = 10^{-7}$, initialize discrete derivative accumulation variables: r_x, r_y

Output: Maximum value of objective function: F_{min}

1: **while** Stop criteria not reached **do**
2:    Fix the x-axis direction, treat variable x as constant;
3:    **repeat**
4:       Calculate the discrete derivative in y direction d_y;
5:       Accumulate discrete derivatives, update r_y;
6:       Calculate and update step size:s_y;
7:       Calculate the next position of the agent in the y direction;
8:       Update agent location;
9:    **until** Find that local optimal solution in the y-axis direction
10:   Fix the y-axis direction, treat variable y as constant;
11:   Repeat the process of finding the local optimal solution on the y-axis;
12: **end while**

Finally, compared with the traditional mathematical optimization process, the CGD has constraints of in vivo computing. The initial location of the agents is limited to a small area, which is the injection point, instead of the whole detection domain. If agents outside the detection domain, new agents will be injected at the initial point to ensure the balance of the total agents.

4   Simulation Results

To validate our model and verify the performance of the proposed CGD algorithm, numerical simulations are complemented through MATLAB. The search range is $(-5\,mm, 5\,mm)$, vessel spacing is $0.1\,mm$, tumor radius is $0.5\,mm$, and the maximum number of iterations is set to 200. Two Manhattan-like functions modified in Sect. 2 are used as objective functions, and the initial position of agents is limited to $-5\,mm \leq x, y \leq -4\,mm$. Two indicators, the iterations K and the effective rate ξ, are introduced to evaluate the efficiency of the CGD algorithm. ξ is the ratio of the number of agents successfully staying in the tumor area to the number of general agents injected. In the simulation, three agents are injected at the same time, We assume that each agent is independent in the computation process. For comparison, brute-force search is also performed in the objective functions, in which the agents wander upward and right randomly at each point.

The initial injection locations of the agents are set as $(-5, -5)$, $(-4, -5)$, $(-4.5, -4.5)$. Any agent staying in the tumor area is recognized to have successfully detected for the tumor. As shown in Fig. 3, the red trajectories represent the CGD algorithm and the black trajectories represent the brute-force.

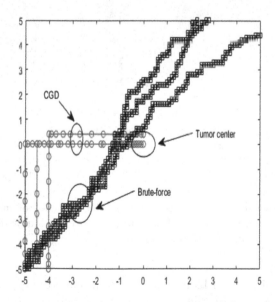

Fig. 3. The red cycles represent the trajectories of the CGD, while the black cubes represent the trajectories of the "brute-force". Agents stay in the tumor area are regarded as a successful homing. (Color figure online)

In the simulation results, because of its randomness, brute-force failed to detect tumor most of the time and the iteration number k of successful detection was approximately 100. And the simulation results of the CGD iterative algorithm are presented in the Table 2. In terms of the simulation results, CGD can detect the tumor location with a higher ξ and fewer K for both different objective functions.

Table 2. K and ξ of the CGD

Objective functions(F)	Initial positions	Iterations(K)	The effective rate(ξ)
Bowl-shaped	$(-5,-5)$	22	50%
	$(-4,-5)$	20	
	$(-4.5,-4.5)$	22	
Plate-shaped	$(-5,-5)$	26	83%
	$(-4,-5)$	25	
	$(-4.5,-4.5)$	26	

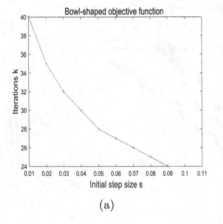 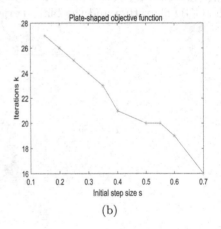

(a)   (b)

Fig. 4. (a) For Bowl-shaped objective function, the influence of initial step size on the number of iterations(K); (b) For Plate-shaped objective function, the influence of initial step size on the number of iterations(K).

Due to the limitation of blood flow direction, nanorobots simply move from one side to the other, quite differ from traditional numerical optimization where computing agents can freely reciprocate in search space. Therefore, the setting of optimal parameters should be different in optimization algorithms for in vivo computing. In our simulation experiments, the selection of the initial step size has a prominent effect on K. As shown in Fig. 4, an increase in initial step size will diminish the number of K. If the initial step size is extremely small, the agents do not have exploration ability at the beginning of the calculation. And if the initial step size exceeds a certain range would result in the poor discovery of the intended tumor location. This indicates that the initial step size is the main parameter of the searching process.

5   Conclusion

This paper introduces Manhattan-like BGFs and exhibits the transformation of objective functions in different dimensions. In addition, we propose a CGD iterative algorithm in a Manhattan-like space to homing tumor and validate its performance through the simulation. This work has enhanced our understanding of BGFs distribution. Further work should focus on more factors of Manhattan-like BGFs formed by tumors. Finally, the tumor homing process should involve external centralized control to ensure the iterations of the agents.

References

1. Friedrich, R.M., et al.: Magnetic particle mapping using magnetoelectric sensors as an imaging modality. Sci. Rep. **9**(1), 1–11 (2019)
2. Thoidingjam, S., Tiku, A.B.: New developments in breast cancer therapy: role of iron oxide nanoparticles. Adv. Nat. Sci. Nanosci. Nanotechnol. **8**(2), 023002 (2017)
3. Bae, Y.H., Park, K.: Targeted drug delivery to tumors: myths, reality and possibility. J. Controlled Release **153**(3), 198 (2011)
4. Chen, Y., Ali, M., Shi, S., Cheong, U.K.: Dispensing-by-learning direct targeting strategy for enhanced tumor sensitization. IEEE Trans. NanoBiosci. **18**(3), 498–509 (2019). https://doi.org/10.1109/TNB.2019.2919132
5. Kasban, H.: A comparative study of medical imaging techniques. Int. J. Inf. Sci. Intell. Syst., **4**, 37–58 (2015)
6. Ali, M., Sharifi, N., McGrath, N., Cree, M.J., Chen, Y.: Self-regulated and coordinated smart tumor homing for complex vascular networks. In: 2020 42nd Annual International Conference of the IEEE Engineering in Medicine Biology Society (EMBC), pp. 378–381 (2020). https://doi.org/10.1109/EMBC44109.2020.9176014
7. A, P.V.: Tumor microenvironmental physiology and its implications for radiation oncology. Seminars in Radiation Oncology **14**(3), 198–206 (2004)
8. Helmlinger, G., Yuan, F., Dellian, M., Jain, R.K.: Interstitial ph and po2 gradients in solid tumors in vivo: high-resolution measurements reveal a lack of correlation. Nat. Med. **3**(2), 177–182 (1997)
9. Baish, J.W., Gazit, Y., Berk, D.A., Nozue, M., Jain, R.K.: Role of tumor vascular architecture in nutrient and drug delivery: an invasion percolation-based network model. Microvasc. Res. **51**(3), 327–346 (1996)
10. Gray, M., Meehan, J., Turnbull, A.K., Martínez-Pérez, C., Argyle, D.J.: The importance of the tumor microenvironment and hypoxia in delivering a precision medicine approach to veterinary oncology. Front. Vet. Sci. **7**, 907 (2020)

11. Dewhirst, M.W., Ong, E.T., Braun, R.D., Smith, B., Klitzman, B., Evans, S.M., Wilson, D.: Quantification of longitudinal tissue po2 gradients in window chamber tumours: impact on tumour hypoxia. Br. J. Cancer **79**(11), 1717–1722 (1999)
12. Srensen, B.S., Horsman, M.R.: Tumor hypoxia: Impact on radiation therapy and molecular pathways. Front. Oncol. **10**, 562 (2020)
13. Cheang, U.K., Meshkati, F., Kim, H., Lee, K., Fu, H.C., Min, J.K.: Versatile micro-robotics using simple modular subunits. Sci. Rep. **6**, 30472 (2016)
14. Agemy, L., Sugahara, K.N., Kotamraju, V.R., Gujraty, K., Ruoslahti, E.: Nanoparticle-induced vascular blockade in human prostate cancer. Blood **116**(15), 2847–2856 (2010)
15. Mcdougall, S.R., Anderson, A., Chaplain, M., Sherratt, J.A.: Mathematical modelling of flow through vascular networks: Implications for tumour-induced angiogenesis and chemotherapy strategies. Bull. Math. Biol. **64**(4), 673–702 (2002)
16. Onose, A., Dumitrescu, B.: Adaptive randomized coordinate descent for solving sparse systems. In: Adaptive Randomized Coordinate Descent for Solving Sparse Systems, IEEE (2014)

Theoretical Basis for Gene Expression Modeling Based on the IEEE 1906.1 Standard

Yesenia Cevallos[1], Tadashi Nakano[2], Luis Tello-Oquendo[1(✉)],
Nishtha Chopra[3], Amin Zadeh Shirazi[4], Deysi Inca[1], and Ivone Santillán[5]

[1] College of Engineering, Universidad Nacional de Chimborazo, Riobamba, Ecuador
lptelloq@ieee.org
[2] Graduate School of Frontier Biosciences, Osaka City University, Osaka, Japan
[3] Queen Mary University of London, London, UK
[4] Centre for Cancer Biology, SA Pathology and the University of South of Australia,
Adelaide, SA 5000, Australia
[5] University of Pavia, Pavia, Italy

Abstract. Molecular communications essentially analyze the transmission of the information at the nano level in cells, the smart devices that constitute our bodies. This emerging field uses traditional communication systems elements and maps them to molecular signaling and communication found inside and outside the body. Hence, molecular communications' fundamental importance denotes the necessity to develop a new technology framework that provides a novel perspective to fight human diseases (the COVID-19 pandemic has highlighted this challenge). Thus, the architecture for molecular communications can be explored from the perspective of computer networks, i.e., the TCP/IP reference model and the basic model of MC can also be represented using Shannon's communication model (i.e., transmitter, communication channel, and receiver). In this field, IEEE impulses the 1906.1 and 1906.1.1 standards that establish definitions, terminology, and a conceptual model for ad hoc network communication at the nanoscale. With these ICT perspectives, we appropriately have analyzed gene expression in eukaryotes organisms as a layered stack (network, link, and physical layer) of a nano communication network. In this biological communication process, the cellular nucleus behaves as the DTE, the ribosomes, and Endoplasmic Reticulum represent the DCE, the Golgi Apparatus represents a border router. The proteins secreted by the cell move through the bloodstream (physical transmission medium) and reaching the receiver (DCE-DTE), which processes the information through ligands and their receptors.

Keywords: Genetic expression · Molecular communication · Stack network modeling · IEEE 1906.1 standard

© ICST Institute for Computer Sciences, Social Informatics and Telecommunications Engineering 2021
Published by Springer Nature Switzerland AG 2021. All Rights Reserved
T. Nakano (Ed.): BICT 2021, LNICST 403, pp. 145–162, 2021.
https://doi.org/10.1007/978-3-030-92163-7_12

1   Introduction

Molecular communications (MCs) essentially analyze the transmission information at the nano level; the cells, the "smart" devices that constitute our bodies, principally communicate fundamentally through molecules' transport and binding. Despite their noisy world and often diffusive transport, nature has devised a way for all of these biomolecules to function in harmony so effectively and reliably, and this is a remarkable realization in robust communication. It's a triumph that scientists must dissect in order to create devices that can communicate with nature in its own language [1].

MCs, in contrast to existing telecommunication paradigms, use molecules as information carriers; sender biological nanomachines (bio-nano machines) encode data on molecules (signal molecules) and release the molecules into the environment. The molecules then travel through the environment to reach the receiver bio-nano machines, where they biochemically react with the molecules to decipher information [2]. In biology, sender-receiver systems abound, with communication systems transmitting data in a variety of ways. Information and communication technology (ICT) offers a quantitative foundation for understanding natural and synthetic genetic processes.

Keeping the preceding ideas in mind, let us investigate a previously unimagined and unfortunate occurrence that has harmed humanity in the area of biological communications. Hundreds of millions of people are infected with viruses each year, but many of them do not have access to vaccines or adequate treatment during and after infection. The COVID-19 pandemic has brought this issue to light, demonstrating how rapidly viruses can spread and affect society as a whole. To fight viral infections and potential future pandemics, new approaches from various disciplines must arise. Over the last ten years, an interdisciplinary field known as molecular communications has emerged [3–5]; it involves bioengineering, nanotechnology, and ICT. This new area applies elements of traditional communication systems to molecular signaling and communication found both within and outside the body. The aim is to create new methods that will aid future medicine (the ICT system offers a novel approach to combating human diseases) [3,6–9]). Characterizing the signaling pathways between cells and infectious disease locations at different levels of the human body is one of these methods [10].

Thus, MCs have been developing, and it bridges communication engineering and networking, molecular biology, and bioengineering areas [11,12]. Much analysis, designs, and proposals in biological systems communications studies have been established from a typical layered network structures paradigm [13–25]. In this area, the Institute of Electrical and Electronics Engineers (IEEE) impulses the 1906.1 and 1906.1.1 standards [26–29]. In its first version, the IEEE 1906.1 standard recommended practice contains a conceptual model and standard terminology for ad hoc network communication at the nanoscale. The new version of the standard i.e., 1906.1.1 establishes a common YANG (Yet Another Next Generation) data model for IEEE 1906.1 nanoscale communication systems.

We introduced a stacked-layer network model to describe gene expression using the fundamentals of communication engineering and networking in MC as a guide, as well as the relevance of series IEEE 1906 standards [30]. The rest of the paper is organized as follows. Section 2 explains gene expression and protein delivery from an ICT paradigm; thus, Sect. 2.1 describes the transcription of DNA (DTE in the transmitter), Sect. 2.2 denotes translation of DNA (DCE in the transmitter), Sect. 2.3 describes to Golgi apparatus (GA) as an internet (border) router, Sect. 2.4 establishes the protein delivery through a communication channel, Sect. 2.5 describes the use of information at the receiver (DCE-DTE). In Sect. 3 (application of the series standards IEEE 1906 to gene expression) is established our proposal, hence Sect. 3.1 indicates the IEEE Standard Data Model for Nanoscale Communication Systems, i.e., IEEE 1906.1 and 1906.1.1, Sect. 3.2 presents the proposal properly as a viewed digital communication networks through gene expression and protein delivery from the IEEE 1906.1 and 1906.1.1 paradigm. Conclusions enclose this paper in Sect. 4.

2 Gene Expression and Protein Delivery Analysis from a Network Layered Paradigm

2.1 DNA Transcription

The basic model of MCs may be described based on Shannon's model of communication (i.e., transmitter, communication channel, and receiver) [20] and as nanoscale communication network is a human-designed system for communicating at or with the nanoscale, using physical principles that are suited to nanoscale systems [26]. Then, from our previous work in [30], we suppose that the cellular nucleus represents the biological transmitter (or a biological Data Terminal Equipment (DTE) that contains the information source) composed of nucleotide blocks called genes (which must be processed intra-cellularly or extra-cellularly). We mainly focus on long-distance cellular communication (exhibited, for example, in the endocrine system, this type of MC is known as long nano range communication) [30–32]. In nature, a gene is a set of nucleotides that stores the information required for accomplishing a specific function (by a protein or RNA) [33]) to be performed at a specific destination; Thus, we hypothesize that a gene's contents can be understood as addressing at the network layer. The transmission of information from the nucleus cell (i.e., from genes) to specific destinations begins with the transcription process in which DNA information is copied into RNA [30].

The DNA molecules are encoded by four distinct values, allowing them to contain digital information (i.e., four nucleotides, then a single nucleotide base will carry two bits of information $[-\log_2(1/4)]$). Nucleotides are monomers of nucleic acids that contain one nitrogenous base, a five-carbon sugar (deoxyribose in DNA and ribose in RNA), and at least one phosphate group (DNA and RNA). Adenine (A), thymine (T), cytosine (C), guanine (G), and uracil (U) are nitrogenous bases (U). The DNA double helix is made up of nucleotides with the bases A, T, C, and G [34], and it keeps its shape since the nitrogenous bases on each strand are complementary [35] (i.e., the affinity of adenine to thymine and that of cytosine to guanine [36]). Since biological information is divided into encapsulated data segments [19, 30], the information in DNA is divided into blocks of nucleotides called genes [37, 38] that have start and termination sequences. Digital information is split into packets in packet-switching networks to speed up transmission. As a result, a digital network packet may be thought of as a gene in a biological transmission network [19, 30].

As transcription begins, the RNA polymerase II (RNAP II enzyme) molecular motor recognizes a region of the promoter region's DNA sequence [39]. When RNAP II starts adding nucleotides to create a complementary messenger RNA (mRNA) chain, it starts with the promoter. After that, RNAP II makes a single-stranded mRNA copy of one of the two DNA strands. The only difference between RNA and DNA is that in this step, RNAP II uses uracil (U) instead of thymine (T) [40, 41].

The enhancer is another essential DNA strand component that controls the amount of protein produced in relation to the amount of mRNA (i.e., the amount of information sent to the receiver). As a consequence, this technique can be interpreted as a sender-side flow control [30]. Flow control is maintained at the data-link layer to ensure that a fast sender does not overwhelm a passive receiver with more messages than can be processed [42–44]. The transcription of one of the DNA strands from the 5'P to the 3'OH of the deoxyribose phosphate backbone was specifically unidirectional [39, 40]. This precision is also present in computer networks, where the less significant bit must be signaled when information is transmitted using serial communication [43].

The halt of transcription is accomplished when an appropriate finalization sequence is recognized by RNAP II [29]. In the primary transcript molecule (i.e., pre-mRNA) occurs maturations such as: (i) Splicing, (ii) Capping, and (iii) Polyadenylation [45]. The information added during capping and polyadenylation may be equivalent to the delimiting data flags used in digital communication systems, for instance, headers and trailers that encapsulate the information (i.e., forms a molecular frame) in the data-link layer in protocol hierarchies in network software [20, 30]. These flags are used for processing and error control [19, 44, 45]; mRNA molecules are stabilized (control and posterior processing) by the previously described maturations, which avoid mRNA degradation by enzymes in the cytosol (intracellular fluid), allowing them to progress to the next step of biological processing [30]. The data added during capping and polyadenylation could be compared to a header and trailer that encapsulates the data.

The mechanical transport of mRNA molecules through the cytosol may be analogous to the transmission of information in wired communications (physical layer task) [19, 30, 31, 36].

2.2 DNA Translation

The transcription process permitted to copy the biological information from DNA to RNA; this is mandatory because DNA molecules cannot leave the cellular nucleus. Hence, at this point, the biological DTE must transmit the information to biological Data Communication Equipment (DCE) through a physical interface (as in conventional communications systems). In our case, supported by [30], we assume that the cytosol may represent this physical interface.

The DCE (codec or modem) is the unit in a digital communication system that is responsible for properly formatting the data transmitted over a communication channel. According to [30], ribosomes and the ER (Endoplasmic Reticulum) serve the biological DCE since genetic information acquires a functional structure (or format when referring to data) through these organelles, which is then released into the biological communication channel and eventually reaches the biological recipient. The "right formatting" of biological data occurs during translation as the information is translated into amino acid chains in order to obtain functionality both within and outside the cell. Thus, the biological DCE processes (codifies) information through translation and provides a specific input sequence (mRNA data) that is associated with a specific output (amino acid sequence); this type of codification process corresponds to traditional codification in digital systems (i.e., a physical layer task) [30, 43, 44]. The natural world has created the opportune intermediate addressing at any point of transcription and translation; hence, the mRNA that leaves the nucleus has an implied contiguous address that is comparable to a data link layer address to enable communication within a direct communication range [19, 30, 43, 44]; as a result, cytosolic ribosomes or those associated with the rough ER (RER) bound the mRNA. The arrival and transfer of biological information through a biological communication medium, which is considered a task at the physical layer, is symbolized by the transmission (movement) of biological information from the nucleus to the ribosomes or ER through the cytosol [19, 30].

Ribosomes are the structures that function as molecular motors. They used a codon scheme (i.e., a triplet of nucleotides) to read the details found in the biological sequence [30, 46]. Transfer RNAs (tRNAs) identify the codons in the mRNA in the ribosome. They have an anticodon sequence that is complementary to a specific codon that is associated with a specific amino acid that attaches to the molecular structure of that tRNA [40].

The signaling of tRNAs triggers protein synthesis in the ribosome. To ensure proper biological information reading, they indicate the process's start and stop codons to the ribosomes [46]. The study of amino acid interactions that shape proteins is critical for understanding evolutionary relationships among species, developing new drugs, and creating synthetic proteins from the perspective of digital communications systems [47]. The start and stop codons can correspond to synchronism signals in the digital communications systems paradigm. A start flag is used to execute synchronous transmissions between the source and destination. In this method of communication, the transmitter sends the data, which the receiver must then collect and process. The stop codon in biological signaling can be analogous to the stop flag used at the destination to signify the end of synchronous communications. Besides, synchronous transmissions are the focus of this investigation since synchronous communications allow for the transmission of large amounts of data; for example, 455 EB of data can be encoded in 1 g of single-stranded DNA [48]. The signals that signal the start and stop of DNA translation enable biological clocks in cells to provide feedback during cellular processes [19, 30].

As the processing of specific amino acids generates proteins, the quantity of information required to specify one amino acid from a set of 20 (the total number of amino acids) unambiguously is 4.3 bits $[-\log_2(1/20)]$; then, using 6 bits to define an amino acid signifies an excess of information; nevertheless, this excess information capacity may explain the genetic code's redundancy [49].

In MCs, the modulation (a physical layer task) is a functionality at the transmitter to alter molecules' properties to represent the information that arrives as a concentration of molecules at the receiver [19]. One mechanism for this modulation is to choose one type of molecule from a set of molecule types, this type of modulation is known as MoSK (Molecular-Shift Keying) [50]. On the other side, as genetic code uses three nucleotides (codon) to represent a specific amino acid and as there are 20 amino acids, it is necessary to sort the four nucleotides into groups of at least 3 to encode all 20 amino acids in 64 possible combinations (i.e., $4^3 = 64$ because $4^2 = 16$ is not sufficient to encode 20 amino acids) [51]. Hence, in our research, we assume that the biological DCE codifies a set of 6 inputs (3 nucleotides or 6 bits due each one is represented by 2 bits) with a set of 64 outputs (to encode the total of 20 amino acids) [30]. Then the DCE in our case may have a 64 MoSK (i.e., there exist 64 different cases to be considered, because, for the MoSK modulation scheme, the information is represented by using different types of molecules, for x bits information per symbol, 2^x types of molecules are needed to transmit [52, 53]).

The protein production processed in the ER (e.g., peptide hormones, such as insulin) is analyzed to achieve long nano-range communication; an addressing scheme similar to the network address in conventional computer networks is needed [19, 30]. Note that many of these proteins have a function outside of the cell, where they enter the receiver in what is known as the long nano range. To obtain these proteins, they must be tagged (i.e., tagging to play a role outside the cell). As a result, the amino acid sequence is bound to a signal recognition particle (SRP), which provides an implicit adjacent address through molecular tagging [54]. This molecular tag functions similarly to a data link layer address in that it facilitates contact within a direct communication range [19, 30].

The primary function of the SRP is to assist the nascent protein in reaching a channel protein in the ER that controls the protein's translocation. The SRP then separates from the protein and returns to the cytosol [30, 54]. Similarly, after processing and control information have been used in automated communication systems, they are discarded [43, 44]. The proteins are folded at this point and acquire the usable three-dimensional structure needed for them to perform their basic biological functions within the ER (equivalent to digital data after processing by the DCE, i.e., getting the necessary format [43, 44]) [30, 54].

In biological systems, information errors can occur during DNA transcription and translation, just as they can in traditional communication systems with errors in the transmission media. Many medical conditions, such as autoimmunity, cancer, and diabetes, are caused by errors in cellular transmission and information exchange [55].

2.3   The Golgi Apparatus (GA) as an Internet (border) Router

The RER transfers to the GA the proteins when they are functional via molecular motors. Since each protein contains an implicit adjacent address (analogous to a data link layer address that aids communication within a direct communication range [19,20,30]), the proteins are routed to the proper inter-cellular destinations (i.e., in long nano range communications with addressing at the network layer); however, the GA decides whether the proteins remain inside the cell [56].

Proteins and their information content transfer from the RER to the GA during this phase. The data is then stored in vesicles that bind to the cis GA face. The protein information is then encoded in new vesicles, and other cellular components needed for protein processing are added. The contents of the new vesicles are deposited into the medial GA face, and new vesicles containing the protein and the elements required for further processing are produced. Finally, the vesicles enter the trans GA face, where they undergo the same process as previously mentioned; as a result, the proteins are incorporated into new vesicles but guided to the endoplasmic membrane, where they will be secreted outside the cell [54]; vesicles may naturally match a molecular frame at the molecular link-layer [19,30].

As previously mentioned, the GA functions are identical to those of a network's boundary router. When encapsulating and unwrapping the information across layers, a router decides if the information stays within the network or leaves it. As a result, the behavior of depositing proteins, forming vesicles, and adding information to determine a protein destination are similar to the processing of protocol data units (PDUs) in a router's layers [30,43,44]. The layout of a layered model decomposes a large-scale system into a series of functionally independent smaller units (i.e. layers) and determines the interactions between the layers [19,43,44]. As a result, the benefits of using a stack of layers include using a data link layer to turn an unstable channel into a line free of transmission errors or reporting unresolved problems to the upper layer [30,57]. As a result, by applying such a model to biological systems, high reliability can be achieved (e.g., in drug delivery) [13–25,30].

2.4   Protein Delivery Through a Communication Channel

The mode of protein delivery depends on the body's destinations and the system's specific requirements (e.g., the endocrine system). Hence, cases in which the proteins secreted by the cell (e.g., hormones of a proteinaceous nature) move through the bloodstream, i.e., the physical transmission medium (active random with drift transport-diffusion with drift) to a target organ (with addressing destination at the network layer) are considered. This type of MC is referred to as intercellular communication (i.e., distances in the range of mm to m) [9,30,32].

Thus, the movement of the molecules in a fluid medium with drift (e.g., the bloodstream) is characterized as follows $f(t) = \sqrt{\frac{\lambda}{2\pi t^3}} \exp\left(-\lambda \frac{(t-\mu)^2}{2\mu^2 t}\right)$, for $t > 0$, where the mean is $\mu = d/v$, the shape parameter is $\lambda = d^2/2D$, the velocity of the fluid medium is $v \geq 0$, the diffusion coefficient is D, and the distance from the transmitter to the receiver is d [9,19,20].

As proteins move to a target destination through the bloodstream, the traffic of biological information from senders to receivers converts the blood into a shared media. This shared link requires media access control (MAC) to divide it among multiple senders and transmits molecular frames from multiple senders without causing interference between molecular frames. One mechanism for MAC is a time-division multiplexing (TDM), in which different senders transmit molecular frames at different times [19,30].

During the transmission of information over communication networks, a number of issues can arise. Biochemical, thermal, and physical noise, interference (which can be regulated by an effective transmission rate), and attenuation are all problems in molecular communication (which depends on the distance traveled and the physical characteristics of the fluid medium) [20,30]. The noise is any distortion that results in degradation of the signal at the receiver mainly due to stochastic nature; in MCs, the noise sources can be a) random propagation (diffusion) noise, b) transmitter emission noise, c) receiver counting/reception noise, d) environment noise such as degradation and/or reaction, e) multiple transmitters [14,20,58–60].

The resulting damage to the signal information can cause latency (i.e., movement delay) calculated as d/v, jitter (i.e., variation in latency) calculated as $D\,d/2\,v^3$, and the loss rate (i.e., the probability that the intended biological receiver does not receive a molecule transmitted by a biological sender) can increase. The loss rate is calculated as $1 - \int_0^T f(t)dt$, assuming that the receiver waits for the time duration T [9,30].

Based on the communicational parameters that denote the problems that can occur in a communication channel with noise, the Shannon theorem is used to determine the maximum biological information transfer speed (channel capacity) as $C = \max\{I(X;Y)\}$ [20,59,61], where $I(X;Y)$ represents the entropy of the mutual information (MI) of X and Y. The information signals at the transmission and reception ends are denoted as X and Y, respectively.

2.5   Use of Information at the Receiver

In the human body (considering Shannon's communication model [20]), the communication of biological information from a transmitter (DTE-DCE) to a receiver (DCE-DTE) is done through the bloodstream (i.e., the communication channel). A target cell, tissue, or organ performs a physiological function (due to a specific type of biological information comparable with network layer addressing [19,30]); then, we focus on the type of receivers located in a long nano range. Therefore, the transmitter sends the information using the data stored in the DNA molecules and at the destination can recognize the target cell, tissue, or organ. In terms of the type of proteinaceous hormone involved, the receiver processes the information received. Here, we briefly describe a case in which this processing is performed through ligands, and their receptors [62].

Signal molecules are received by receptors within the receiver in nature; the receiver is made up of a chemical detector that senses the concentration of molecules at a given sample time and demodulates the signal [20,63,64]. As a result, these protein structures can be thought of as antennas for receiving signals. Receptors are protein structures that have the ability to bind to complex ligand structures. Intermolecular forces such as ionic bonds, hydrogen bonds, and van der Waals forces are used to tie molecules together. In most cases, docking (association) is reversible (dissociation). The chemical conformation of a receptor is altered when a ligand binds to it, and the propensity to bind is known as affinity. This method of communication resembles a lock and key system in which only the "key" receiver can detect, read, and interpret the information; other receivers can detect but not process the information, implying process specificity [30]. A receptor's functional state is determined by its conformational state. During the detection process, almost all ligand structures in nature catch and remove the propagation environment's information particles [14,20].

The membrane protein acts as a transducer in the biological process, decoding the received signal (i.e., performing typical DCE tasks at the receiver) and causing multiple reactions within the target cell, tissue, or body organ (i.e., accomplishing the typical DTE tasks at the receiver) [65]. This activity is similar to the work performed at the receiver end of digital communication systems to process data that will be useful at the destination [66,67]. The biological data is transmitted to other organelles after the target cell receives it, using an implicit adjacent address similar to the data link layer address used to enable communication within a direct communication range [19,30]. The biological message that has been received is physically transmitted to the target cell.

3 Application of the Series Standards IEEE 1906 to Gene Expression and Protein Delivery

3.1 IEEE Standard Data Model for Nanoscale Communication Systems IEEE 1906.1 and 1906.1.1

In its first version, the IEEE 1906.1 standard recommended practice (approved in 2015 [27]) contains a conceptual model and standard terminology for ad hoc network communication at the nanoscale. More specifically, this recommended practice contains: a) the definition of nanoscale communication networking; b) the conceptual model for ad hoc nanoscale communication networking; c) the common terminology for nanoscale communication networking, including 1) the definition of a nanoscale communication channel highlighting the fundamental differences from a macroscale channel; 2) abstract nanoscale communication channel interfaces with nanoscale systems; 3) performance metrics common to ad hoc nanoscale communication networks; 4) the mapping between nanoscale and traditional communication networks, including necessary high-level components such as a map of significant components: coding and packets, addressing, routing, localization, layering, and reliability.

According to [27], a common abstract model enables theoretical progress to proceed from different disciplines with a common language. As the industry becomes more interested in the commercial integration of the technology, this structure serves as a recommended practice for additional nanoscale communication networking standards. The biomedical industry needs nanoscale communication standards to create breakthrough diagnostic and treatment methods. Technical discussions and establishing standards in nanoscale communications are impaired by lacking a common conceptual model and standard nomenclature.

In [28], a new version of the standard was developed, i.e., 1906.1.1 (approved in 2020 by Internet Engineering Task Force, Request for Comment 7950) in which IEEE establishes a common YANG (Yet Another Next Generation) data model for IEEE 1906.1 nanoscale communication systems.

In accordance with IEEE Std 1906.1-2015, this data model consists of a series of YANG modules that describe nanoscale communication systems and their associated physical quantities (a common framework for all nanoscale communication technologies). The model depicts physics that are only found at the nanoscale. The physics are non-standard, as specified by IEEE Standard 1906.1-2015. For remote operation and study of nanoscale communication systems, the model specifies remote configuration and management. The model for datastores and repositories of nanoscale communication experimental data defines a self-describing data structure, allowing for a shared interpretation of data from a wide range of nanoscale communication media and technologies.

The YANG data model defines a standard network management and configuration data model for nanoscale communication systems. In doing so, it fulfills several purposes:

- Enforces requirements to conform to IEEE Std 1906.1(TM)-2015.
- Describes nanoscale communication systems.
- Represents the fundamental physics impacting IEEE 1906.1 systems.
- Defines configuration and management for simulation and analysis.
- Defines a self-describing data structure used in repositories of nanoscale communication experimental data.

IEEE 1906.1 systems and simulations can be easily understood and used thanks to a common network, management, and configuration data model. To ensure that systems and simulations comply with IEEE Std 1906.1-2015, a standard data model is required. IEEE 1906.1 systems need a standard data model to act as human and machine-readable documentation. Since nanoscale physics interacts directly with small-scale communication systems, a data model that represents fundamental physics is needed. To compare IEEE 1906.1

systems accurately and equally, a standard data model is needed. For experimental data from small-scale communication systems to be relevant, repositories need detailed and reliable documentation. This standard data model tackles this need by providing a self-describing data model.

3.2 Viewing Digital Communication Networks Through Gene Expression and Protein Delivery from the IEEE 1906.1 and 1906.1.1 Paradigm

The following Tables (1, 2, 3 and 4) provide an analogy of the IEEE Standards 1906.1 and 1906.1.1 for both Molecular Communications and Gene Expression and Protein Delivery. We do not apply specific metrics (defined in both standards 1906.1 and 1906.1.1) to characterize gene expression due we describe this biological process from a general perspective without particularizing the specific case of a particular type of protein.

It is worth noting that nanoscale communication networks shall describe their physical layer by denoting: transmitter, receiver, message, medium, components that have a dimension from 1 nm to 100 nm, the communication physics suited to the nanoscale, message carrier, motion, field, perturbation, and specificity (this is a definition from Sections 3 and 4 in IEEE 1906.1 and Section 3 in IEEE 1906.1.1). On the other hand, Gene expression and protein delivery as nano communication networks seem inertial described (as seen in the whole document) in the standard IEEE 1906.1 and IEEE 1906.1.1 as indicated previously.

Table 1. Introduction section (in IEEE 1906.1)

Molecular communications	Gene expression and protein delivery
(a) In the molecular nanoscale communication embodiment, it is assumed that molecules move into the medium following the omni-directional Fick's law (molecular concentration as a function of distance and time)	(a) $f(t) = \sqrt{\frac{\lambda}{2\pi t^3}} \exp\left(-\lambda \frac{(t-\mu)^2}{2\mu^2 t}\right), \quad t > 0,$ where the mean is $\mu = d/v$, the shape parameter is $\lambda = d^2/2D$, the velocity of the fluid medium is $v \geq 0$, the diffusion coefficient is D, and the distance from the transmitter to the receiver is d [9,19,20]
(b) Knowing the number of molecules released for each pulse	(b) 64 MoSK [30,51–53]
(c) Evaluation of the propagation delay	(c) latency $= d/v$, jitter $= D\,d/2\,v^3$ [9,30]
(d) Estimation of the maximum channel capacity when a concentration-based receiver is used	(d) $C = \max\{I(X;Y)\}$, where $I(X;Y)$ represents the entropy of the mutual information (MI) of X and Y. The information signals at the transmission and reception ends are denoted as X and Y, respectively

Table 2. Definition section (Sections 3 and 4 in IEEE 1906.1 and Section 3 in IEEE 1906.1.1)

Molecular communications	Gene expression and protein delivery
a) A nanoscale communication network is a human-designed system for communicating at or with the nanoscale, using physical principles suited to nanoscale systems b) Communication is the act of conveying a message from a transmitting party to a receiving party. This includes the components (a required element of the framework that provides a communication service in a network) of message, transmitter, receiver, medium, and message carriers. At least one of these components must have nanoscale dimensions. The standard has been defined as nanoscale dimensions between 1 nm to 100 nm c) Active network: A network composed of packets flowing through a telecommunications pathway that dynamically modifies its operation. A packet of information encapsulated to be transported through a communication network d) Message: The information to be conveyed is known to the transmitting party, interfacing with a receiver and unknown but recognizable to the receiving party. The message relates approximately to a classical frame, packet, or protocol data unit (PDU) e) Message carrier: A physical entity that conveys a message across the medium. Message Carrier relates to a wave (the characteristic of a wave that encodes information) f) Transmitter: A device used to convey a message to a receiver g) Receiver: A device used to collect messages from a transmitter h) Receptor: A component that receives signals i) Relay: A component that facilitates communication between a transmitter and the receiver. This network element's motivation is to enable messages to travel longer distances and increase the likelihood of message deliverability. This might be thought of as a form of signal amplification (Section 5.3.2 IEEE 1906.1) j) Medium: The environment connecting the transmitter and receiver, which can include gas, gel, or liquid k) Specificity (sometimes called the actual negative rate) measures the proportion of negatives that are correctly identified (e.g., the percentage of Message Carriers not addressed to an intended target node that is not accepted by the intended target node) l) Sensitivity (also called the true positive rate-the correct classification of a signal, or the recall rate in some fields) measures the proportion of true positives which are correctly identified (e.g., the percentage of Message Carriers addressed to an intended target node that are recognized and accepted by the correct intended target node). Specificity and Sensitivity are widely defined in Section 6.12 in IEEE 1906.1 m) The signaling molecules and cell surface receptor, acting as a message and receiver respectively	a), b) The architecture for MC may be discussed from a computer networks perspective, i.e., the TCP/IP reference model and the basic model of MC may be described based on Shannon's model of communication (i.e., transmitter, communication channel, and receiver) [13,16,25]. Thus, we have analyzed gene expression in eukaryotes organisms (which dimensions are in nanoscale range) from this ICT perspective.In this biological communication process, the cellular nucleus behaves as the DTE, the ribosomes and ER represent the DCE, the GA represents a border router. The proteins secreted by the cell move through the bloodstream (physical transmission medium) and reaching the receiver (DCE-DTE), which processes the information through ligands and their receptors [30,62]. c) Supported by [30] is possible to hypothesize that a packet in a digital network may be analogous to a gene in a biological communication network. These genes are "encapsulated and unencapsulated" through their codification (64 MoSK) and with structures (as start and termination sequences and with capping, and polyadenylation, in transcription); additionally, the genes processing and subsequent transporting through GA to the appropriate destinations correspond to operation encapsulating and unencapsulated of this biological information [19,20,43,44,56] d), e), f), g), h), i), j) To perform a biological role, proteins are formed in the cell and excreted. The biological information source (signal molecules) in eukaryotic cells is the nucleus, which must be changed through intracellular communication to reach an appropriate cellular or extracellular destination. This procedure entails the transmission of information through a biological pathway, followed by the performance of a biological function. The Golgi apparatus is involved in the transmission of proteins (specifically peptide hormones) to a target organ via the bloodstream, which is analogous to the digital communication mechanism in which a transmitter in one network sends data to a destination system in another network through a router (relay component) [30]. Hence, having in mind Shannon's model of communication [20], the communication of biological information from a transmitter (DTE-DCE) to receptors in the receiver (DCE-DTE) is done through the bloodstream (i.e., the communication channel) and in this way, a target cell, tissue or organ performs a physiological function (due to a specific type of biological information which is comparable with network layer addressing [19,30]); then, we focus in the type of receivers that are located in a long nanorange (in which GA determines whether the information remains or leaves the cell, processing biological information as a conventional PDU). Therefore, the transmitter sends the information by using the data stored in the DNA molecules and at destination can recognize the target cell, tissue or organ [62]. In every stage of gene expression and delivery protein, mother nature has established the opportune intermediate addressing (data link layer addressing to facilitate communication within a direct range of communication) and physical layer functions also are executed to transport physically the information in this end-to-end communications nanonetwork [19,30,43,44] k) MCs systems and typical communication systems can encounter problems during the transmission of information via communication channels. Specifically, in MCs, these problems include biochemical, thermal and physical noise, interference, and attenuation [20,30] l), m) Signal molecules are received by receptors within the receiver in nature; the receiver is made up of a chemical detector that detects the concentration of molecules at a given sample time and demodulates the signal [20,63,64]. As a result, these protein structures can be thought of as antennas for receiving signals. Receptors are protein structures that have the ability to bind to various ligand structures. Intermolecular forces such as ionic bonds, hydrogen bonds, and van der Waals forces are used to tie molecules together. In most cases, docking (association) is reversible (dissociation). The chemical conformation of a receptor is altered when a ligand binds to it, and the propensity to bind is referred to as affinity. A receptor's functional state is determined by its conformational state. During the detection process, almost all ligand structures in nature catch and extract information particles from the propagation area [14,20]

Table 3. Framework section (Section 5.2 in IEEE 1906.1 and Section 3.1 in IEEE 1906.1.1)

Molecular communications	Gene expression and protein delivery
a) Component 0: Message Carrier. The Message Carrier provides the service of transporting the message. The message carrier may be either particle or wave. Like quantum mechanics, the message carrier may also be a simultaneous combination of both particle and wave. A molecular structure may encode information transported by the Message Carrier from a transmitter to a receiver. Wave-like changes in message concentration may also encode information. In IEEE 1906.1.1 (Section 5.3.3.3.2) is defined some core message carrier specifications (with a label name) which include molecular motors.	a) Gene expression (transcription and translation of DNA) is how information carried by deoxyribonucleic acid (DNA) is transformed into proteins. The transmission of information from the nucleus cell (i.e., from genes) to specific destinations begins with the transcription process in which DNA information is copied into RNA [30]. The DNA molecules contain digital information due it is encoded by four discrete values (four nucleotides, the quantity of information carried by a single nucleotide base is 2 bits $[-\log_2(1/4)]$). To begin transcription, the molecular motor RNA polymerase II (the RNAP II enzyme) recognizes a region of the DNA sequence called the promoter region [39]. The halt of transcription is accomplished when an appropriate finalization sequence is recognized by RNAP II [29]. In translation, fundamentally, ribosomes, which serve as molecular motors, read the information in the biological sequence using a codon system [30,46]. In the ribosome, the codons in the mRNA are recognized by transfer RNAs (tRNAs) that possess an anticodon associated with a unique amino acid that binds specifically to the molecular structure of that tRNA [40]
b) Component 1: Motion. Defines the movement capability for Component 0 (Message Carrier). The Motion Component provides the service of movement for the Message Carrier (in any direction) caused by force or thrust applied to the Message Carrier. Motion provides the necessary potential to transport information through a communication channel. Message Carriers can be active, generating their motion, or passive, being propagated by the Media. Examples include Molecules diffusing through fluids, Brownian motion, self-propelled motion. Motion (Component 1) relates approximately to the classical physical layer (wave propagation). In IEEE 1906.1.1 (Section 5.3.3.3.2) are defined some motion specifications (with a label name) which include movement through diffusion in this component	b) We focus on the protein delivery to destinations in the endocrine system (e.g., hormones of a proteinaceous nature) that move through the bloodstream (physical transmission or diffusion medium) [9,30,32]
c) Component 2: Field. Defines organized movement of Component 1 (Motion). The Field Component provides the service of an organized motion for Message Carriers. It can be thought of as a virtual waveguide in communications. The Field may be implemented internally or externally relative to the medium. Examples include an internal implementation includes swarm motion or flocking behavior; external implementations are non-turbulent fluid flow, EM field, chemical gradient released to guide movement of bacteria, molecular motors guided by microtubules. Field (Component 2) relates approximately to the classical data link and network layers (ensuring node-to-node information flow). In IEEE 1906.1.1 (Section 5.3.3.3.2) are defined some field specifications (with a label name) which include concentration-gradient in this component	c) In a random walk with drift, information molecules may undergo a directional drift that continuously propagates molecules in the drift's direction (concentration-gradient). An example of this class of MCs is found in our body; cells in the body secrete hormonal substances that circulate with the bloodstream flow and propagate to distant target cells distributed throughout the body. This type of communication also represents the active mode of MC [9,30,68]
	d) We consider that the ribosomes and ER represent the biological DCE because, through these organelles, the genetic information acquires a functional structure (or format when referring to data). This biological DCE processes (codifies) information via translation and provides a specific input sequence (data in mRNA) that is associated with a specific output (sequence of amino acids). Hence, in our research, the biological DCE codifies a set of 6 inputs (3 nucleotides or 6 bits due each one is represented by 2 bits) with a set of 64 outputs (to encode the total of 20 amino acids) [30]. Then the DCE in our case may have a 64 MoSK (i.e., there exist 64 different cases to be considered, because, for the MoSK modulation scheme, the information is represented by using different types of molecules (molecular structure) for n bits information per symbol, 2^n types of molecules are needed to transmit [52,53])
d) Component 3: Perturbation Defines the signal transported by Component 0 (Message Carrier). The Perturbation Component provides the service of varying Message Carriers as needed to represent a signal. This may be thought of as modulation (signal impression). Examples include signals based on the number of received message carriers, controlled dense versus-sparse concentrations of molecules, simple on-versus-off flow of signal molecules, using different types of message carriers, modifying the conformation of molecules (e.g., deoxyribonucleic acid [DNA]) to represent multiple states. In IEEE 1906.1.1 (Section 5.3.3.3.2) are defined some perturbation specifications (with a label name) which include molecular-structure in this component. Perturbation (Component 3) relates approximately to classical modulation at the physical layer	e) Signal molecules are received through protein structures called receptors within the receiver in nature (in the long nano range, an addressing network layer addressing [19,30] is required); the receiver consists of a chemical detector, which will sense the concentration of molecules at a particular sample time and demodulate the signal [20,63,64]. As a result, these protein structures can be thought of as antennas for receiving signals. Receptors are protein structures that have the ability to bind to various ligand structures. Intermolecular forces such as ionic bonds, hydrogen bonds, and van der Waals forces are used to tie molecules together. In most cases, docking (association) is reversible (dissociation). The chemical conformation of a receptor is altered when a ligand binds to it, and the propensity to bind is referred to as affinity. This method of communication is similar to a lock and key system, in which the information is detected, read, and interpreted by a receiver with the "key"; other receivers can detect the information, but they are unable to process it, indicating the process's uniqueness [30]
e) Component 4: Specificity Defines targeted reception of Component 3 (Perturbation). The Specificity Component provides the service of sensing or reception of a message carrier by a target. This can be mapped to addressing in classical communication systems. Examples include the shape or affinity of a molecule to a particular target, complementary DNA for hybridization. In IEEE 1906.1.1 (Section 5.3.3.3.2) are defined some specificity specifications (with a label name) which include receptor-sensitivity in this component. Specificity (Component 4) relates approximately to classical addressing at the data link layer	A receptor's functional state is determined by its conformational state. During the detection process, almost all ligand structures in nature catch and extract information particles from the propagation area [14,20]. The membrane protein acts as a transducer in the biological process mentioned, decoding the obtained signal (i.e. performing standard DCE tasks at the receiver) and causing multiple reactions within the target cell, tissue, or organ of the body (i.e. accomplishing the typical DTE tasks at receiver) [65]. This activity is similar to the work performed at the receiver end of digital communication systems to process data that will be useful at the destination [66,67]. When the biological data is received by the target cell, it is transmitted to other organelles through an implicit adjacent address, which is similar to the data link layer address used to enable contact within a direct range of communication [19,30]. The biological message that has been received is physically transmitted to the target cell

Table 4. Framework components section (Section 5.3.5 in IEEE 1906.1 also supported in IEEE 1906.1.1)

Molecular communications	Gene expression and protein delivery
a) Message-to-Message Carrier (encoding) b) Message Carrier-to-Motion (range of motion) c) Motion-to-Field (controlled motion) d) Field-to-Perturbation (rapid control of field) e) Perturbation-to-Specificity (ability to dynamically change Specifity to encode a message) f) Specificity-to-Message Carrier (Message Carrier and binding capability) g) Message Carrier-to-Receiver (decoding)	a), b), c), d), e), f), g) Having in mind that the intracellular distance is in the range <100 nm and that a typical long nano range is in the range of mm to m [20,31,32,69], and that the functions of biological DTEs and DCEs have been described in this document

4   Conclusions

Molecular communications essentially analyze the transmission of the information at the nano-level in cells, the "smart" devices that constitute our bodies. This new emerging area uses classical communication systems elements and maps it to molecular signaling and communication found inside and outside the body. Hence, molecular communications' fundamental importance denotes the necessity to develop a new technology framework that provides a novel perspective to fight human diseases (the COVID-19 pandemic has highlighted this challenge). As a result, the architecture for molecular communications can be explored from the perspective of computer networks, i.e., the TCP/IP reference model and the basic model of MC can also be represented using Shannon's communication model (i.e., transmitter, communication channel, and receiver). In this area, IEEE impulses the 1906.1 and 1906.1.1 standards that establish definitions, terminology, and a conceptual model for nano communication networks. With these ICT perspectives, we have analyzed gene expression and protein delivery as nanoscale communications networks that describe functions at the physical layer (real communication through a physical medium), link layer (to facilitate communication within a direct range of communication), and network layer (the contents of a gene can be understood as addressing that performs a physiological function at target destination in a long nano communication) by denoting: transmitter, receiver, message, medium, and communication components (i.e., message carrier, motion, field, perturbation, and specificity) from the mentioned standards perspective.

Acknowledgment. The authors would like to thank the financial support of the Ecuadorian Corporation for the Development of Research and the Academy (RED CEDIA) in the development of this work through the Divulga Ciencia program.

References

1. Pilkiewicz, K.R., Rana, P., Mayo, M.L., Ghosh, P.: Molecular communication and cellular signaling from an information-theory perspective. In: Nanoscale Networking and Communications Handbook, p. 235 (2019)
2. Menendez, D.B., Senthivel, V.R., Isalan, M.: Sender-receiver systems and applying information theory for quantitative synthetic biology. Curr. Opin. Biotechnol. **31**, 101–107 (2015)
3. Felicetti, L., Femminella, M., Reali, G.: Modeling approaches for simulating molecular communications. In: Shen, X., Lin, X., Zhang, K. (eds.) Encyclopedia of Wireless Networks, pp. 904–910. Springer, Cham (2020). https://doi.org/10.1007/978-3-319-78262-1_232
4. Cevallos, Y., et al.: On the efficient digital code representation in DNA-based data storage. In: Proceedings of the 7th ACM International Conference on Nanoscale Computing and Communication, pp. 1–7 (2020)
5. Gohari, A., Mirmohseni, M., Nasiri-Kenari, M.: Information theory of molecular communication: directions and challenges. IEEE Trans. Mol. Biol. Multi Scale Commun. **2**(2), 120–142 (2016)
6. Cevallos, Y., Tello-Oquendo, L., Inca, D., Ghose, D., Shirazi, A.Z., Gomez, G.A.: Health applications based on molecular communications: a brief review. In: 2019 IEEE International Conference on E-health Networking, Application and Services (HealthCom), pp. 1–6. IEEE (2019)
7. Zadeh Shirazi, A., et al.: The application of deep convolutional neural networks to brain cancer images: a survey. J. Personalized Med. **10**(4), 224 (2020)
8. Koca, C., Civas, M., Sahin, S.M., Ergonul, O., Akan, O.B.: Molecular communication theoretical modeling and analysis of SARS-CoV-2 transmission in human respiratory system. arXiv preprint arXiv:2011.05154 (2020)
9. Nakano, T., Moore, M.J., Wei, F., Vasilakos, A.V., Shuai, J.: Molecular communication and networking: opportunities and challenges. IEEE Trans. Nanobiosci. **11**(2), 135–148 (2012)
10. Wang, J., Peng, M., Liu, Y., Liu, X., Daneshmand, M.: Performance analysis of signal detection for amplify-and-forward relay in diffusion-based molecular communication systems. IEEE Internet Things J. **7**(2), 1401–1412 (2019)
11. Barros, M.T., et al.: Molecular communications in viral infections research: modelling, experimental data and future directions. arXiv preprint arXiv:2011.00002 (2020)
12. Schurwanz, M., Hoeher, P.A., Bhattacharjee, S., Damrath, M., Stratmann, L., Dressler, F.: Infectious disease transmission via aerosol propagation from a molecular communication perspective: Shannon meets coronavirus. arXiv preprint arXiv:2011.00290 (2020)
13. Dong, M., Li, W., Xu, X.: Evaluation and modeling of HIV based on communication theory in biological systems. Infect. Genet. Evol. **46**, 241–247 (2016)
14. Farsad, N., Yilmaz, H.B., Eckford, A., Chae, C.-B., Guo, W.: A comprehensive survey of recent advancements in molecular communication. IEEE Commun. Surv. Tutorials **18**(3), 1887–1919 (2016)
15. Ali, N.A., Abu-Elkheir, M.: Internet of nano-things healthcare applications: requirements, opportunities, and challenges. In: 2015 IEEE 11th International Conference on Wireless and Mobile Computing, Networking and Communications (WiMob), pp. 9–14. IEEE (2015)

16. Dubey, A., Tandon, S., Seth, A.: Design of a molecular communication framework for nanomachines. In: 2012 Fourth International Conference on Communication Systems and Networks (COMSNETS), pp. 1–2. IEEE (2012)
17. Huang, J.-T., Lai, H.-Y., Lee, Y.-C., Lee, C.-H., Yeh, P.-C.: Distance estimation in concentration-based molecular communications. In: 2013 IEEE Global Communications Conference (GLOBECOM), pp. 2587–2591. IEEE (2013)
18. Abbasi, Q.H., et al.: Nano-communication for biomedical applications: a review on the state-of-the-art from physical layers to novel networking concepts. IEEE Access 4, 3920–3935 (2016)
19. Nakano, T., Suda, T., Okaie, Y., Moore, M.J., Vasilakos, A.V.: Molecular communication among biological nanomachines: a layered architecture and research issues. IEEE Trans. Nanobiosci. 13(3), 169–197 (2014)
20. Nakano, T., Eckford, A.W., Haraguchi, T.: Molecular Communication. Cambridge University Press, Cambridge (2013)
21. Shirazi, A.Z., Mazinani, S.M., Eghbal, S.K.: Protocol stack for nano networks. In: 2012 International Symposium on Computer, Consumer and Control, pp. 849–853. IEEE (2012)
22. Wysocki, B.J., Martin, T.M., Wysocki, T.A., Pannier, A.K.: Modeling nonviral gene delivery as a macro-to-nano communication system. Nano Commun. Networks 4(1), 14–22 (2013)
23. Chude-Okonkwo, U., Malekian, R., Maharaj, B.T.: Internet of Things for advanced targeted nanomedical applications. In: Advanced Targeted Nanomedicine. NN, pp. 113–124. Springer, Cham (2019). https://doi.org/10.1007/978-3-030-11003-1_6
24. Walsh, F., et al.: Hybrid DNA and enzyme based computing for address encoding, link switching and error correction in molecular communication. In: Cheng, M. (ed.) NanoNet 2008. LNICST, vol. 3, pp. 28–38. Springer, Heidelberg (2009). https://doi.org/10.1007/978-3-642-02427-6_7
25. El-taweel, A., Abd El-atty, S.M., El-Rabaie, S.: Efficient molecular communication protocol based on mobile ad-hoc nanonetwork. Menoufia J. Electron. Eng. Res. 26(2), 427–443 (2017)
26. IEEE: IEEE recommended practice for nanoscale and molecular communication framework, IEEE Std 1906.1-2015, pp. 1–64 (2016)
27. IEEE: IEEE standard data model for nanoscale communication systems, IEEE Std 1906.1.1-2020, pp. 1–142 (2020)
28. Bush, S.F.: Interoperable nanoscale communication [future directions]. IEEE Consum. Electron. Mag. 6(2), 39–47 (2017)
29. Bush, S.F., Paluh, J.L., Piro, G., Rao, V., Prasad, R.V., Eckford, A.: Defining communication at the bottom. IEEE Trans. Mol. Biol. Multi Scale Commun. 1(1), 90–96 (2015)
30. Cevallos, Y., Molina, L., Santillán, A., De Rango, F., Rushdi, A., Alonso, J.B.: A digital communication analysis of gene expression of proteins in biological systems: a layered network model view. Cogn. Comput. 9(1), 43–67 (2017)
31. Giné, L.P., Akyildiz, I.F.: Molecular communication options for long range nanonetworks. Comput. Networks 53(16), 2753–2766 (2009)
32. Akyildiz, I.F., Pierobon, M., Balasubramaniam, S., Koucheryavy, Y.: The internet of bio-nano things. IEEE Commun. Mag. 53(3), 32–40 (2015)
33. Meneely, P.: Genetic Analysis: Genes, Genomes, and Networks in Eukaryotes. Oxford University Press, Oxford (2020). https://books.google.com.ec/books?id=SgjZDwAAQBAJ

34. Beauchaine, T., Crowell, S.: The Oxford Handbook of Emotion Dysregulation. Oxford Library of Psychology Series. Oxford University Press, Oxford (2020). https://books.google.com.ec/books?id=UdbODwAAQBAJ

35. Washington, C., Leaver, D.: Principles and Practice of Radiation Therapy. Elsevier Health Sciences, New York (2015). https://books.google.com.ec/books?id=N-h1BwAAQBAJ

36. Nurunnabi, M., McCarthy, J.: Biomedical Applications of Graphene and 2D Nanomaterials. Micro and Nano Technologies. Elsevier Science, New York (2019). https://books.google.com.ec/books?id=BruPDwAAQBAJ

37. Vaz, M., Raj, T., Anura, K.: Guyton & Hall Textbook of Medical Physiology - E-Book: A South Asian Edition. Elsevier Health Sciences, New York (2016). https://books.google.com.ec/books?id=J_HQDwAAQBAJ

38. Barber, J., Rostron, C.: Pharmaceutical Chemistry. Integrated Foundations of Pharmacy. OUP Oxford (2013). https://books.google.com.ec/books?id=do6cAQAAQBAJ

39. Mougios, V.: Exercise Biochemistry. Human Kinetics, Incorporated (2019). https://books.google.com.ec/books?id=Gt-DDwAAQBAJ

40. Bhutani, S.: Chemistry of Biomolecules, 2nd edn. CRC Press, Boca Raton (2019). https://books.google.com.ec/books?id=TO-yDwAAQBAJ

41. Shen, C.: Diagnostic Molecular Biology. Elsevier Science, New York (2019). https://books.google.com.ec/books?id=KGGdBgAAQBAJ

42. Nakano, T., Okaie, Y., Vasilakos, A.V.: Transmission rate control for molecular communication among biological nanomachines. IEEE J. Sel. Areas Commun. **31**(12), 835–846 (2013)

43. Comer, D.: Computer Networks and Internets. Always Learning. Pearson, London (2015). https://books.google.com.ec/books?id=PBhAngEACAAJ

44. Kurose, J., Ross, K.: Computer Networking: A Top-Down Approach, Global Edition. Pearson Education Limited, London (2017). https://books.google.com.ec/books?id=IUh1DQAAQBAJ

45. Reece, J.B., Meyers, N., Urry, L.A., Cain, M.L., Wasserman, S.A., Minorsky, P.V.: Campbell Biology Australian and New Zealand Edition. Pearson Higher Education AU, vol. 10 (2015)

46. Kampourakis, K.: Making Sense of Genes. Cambridge University Press, Cambridge (2017). https://books.google.com.ec/books?id=FV55DwAAQBAJ

47. Chahibi, Y.: Molecular communication for drug delivery systems: a survey. Nano Commun. Networks **11**, 90–102 (2017)

48. De Silva, P.Y., Ganegoda, G.U.: New trends of digital data storage in DNA. BioMed Res. Int. **2016** (2016)

49. Konieczny, L., Roterman-Konieczna, I., Spólnik, P.: Systems Biology: Functional Strategies of Living Organisms. https://books.google.com.ec/books?id=g4y4BAAAQBAJ

50. Chen, X., Huang, Y., Yang, L.-L., Wen, M.: Generalized molecular-shift keying (gmosk): principles and performance analysis. IEEE Trans. Mol. Biol. Multi Scale Commun. **6**(3), 168–183 (2020)

51. Alberts, B.: How cells read the genome: from DNA to protein. In: Molecular Biology of the Cell (2002)

52. Kim, N.-R., Eckford, A.W., Chae, C.-B.: Symbol interval optimization for molecular communication with drift. IEEE Trans. Nanobiosci. **13**(3), 223–229 (2014)

53. Lu, Y.: Error correction codes for molecular communication systems, Ph.D. dissertation, University of Warwick (2016)

54. Plopper, G., Ivankovic, D.: Principles of Cell Biology. Jones & Bartlett Learning, Burlington (2020). https://books.google.com.ec/books?id=mqDHDwAAQBAJ
55. Bush, S.F.: Nanoscale Communication Networks. Artech House, Norwood (2010)
56. Russell, P., Hertz, P., McMillan, B., Benington, J.: Biology: The Dynamic Science. Cengage Learning, Boston (2020). https://books.google.com.ec/books?id=bAfFDwAAQBAJ
57. Chiang, M., Low, S.H., Calderbank, A.R., Doyle, J.C.: Layering as optimization decomposition: a mathematical theory of network architectures. Proc. IEEE **95**(1), 255–312 (2007)
58. Haselmayr, W., Varshney, N., Asyhari, A.T., Springer, A., Guo, W.: On the impact of transposition errors in diffusion-based channels. IEEE Trans. Commun. **67**(1), 364–374 (2018)
59. Cevallos, Y., Tello-Oquendo, T., Inca, D., Palacios, C., Rentería, L.: Genetic expression in biological systems: a digital communication perspective. Open Bioinform. J. **12**(1) (2019)
60. Hong, L., Chen, W., Liu, F.: Cooperative molecular communication for nanonetwork. In: 2014 Sixth International Conference on Ubiquitous and Future Networks (ICUFN), pp. 369–370. IEEE (2014)
61. Raut, P., Sarwade, N.: Establishing a molecular communication channel for nano networks. Int. J. VLSI Des. Commun. Syst. (VLSICS) **4** (2013)
62. Najman, S.: Current Frontiers and Perspectives in Cell Biology. IntechOpen, London (2012). https://books.google.com.ec/books?id=idCPDwAAQBAJ
63. Prep, K.: AP Biology Prep Plus 2020 & 2021: 3 Practice Tests + Study Plans + Review + Online. Kaplan Test Prep. Kaplan Publishing, New York (2020). https://books.google.com.ec/books?id=PM_XDwAAQBAJ
64. Li, B., Sun, M., Wang, S., Guo, W., Zhao, C.: Low-complexity noncoherent signal detection for nanoscale molecular communications. IEEE Trans. Nanobiosci. **15**(1), 3–10 (2015)
65. Pack, P.: Cliffsnotes AP Biology 2021 Exam. Houghton Mifflin Harcourt Publishing Company, Boston (2020). https://books.google.com.ec/books?id=ZAa9DwAAQBAJ
66. Raut, P., Sarwade, N.: Study of environmental effects on the connectivity of molecular communication based internet of nano things. In: International Conference on Wireless Communications, Signal Processing and Networking (WiSPNET), pp. 1123–1128. IEEE (2016)
67. Li, Z.P., Zhang, J., Zhang, T.C.: Concentration aware routing protocol in molecular communication nanonetworks. Appl. Mech. Mater. **556**, 5024–5027 (2014)
68. Arjmandi, H., Zoofaghari, M., Rouzegar, S.V., Veletić, M., Balasingham, I.: On mathematical analysis of active drug transport coupled with flow-induced diffusion in blood vessels. IEEE Trans. NanoBiosci. **20**(1), 105–115 (2020)
69. Furubayashi, T., Nakano, T., Eckford, A., Okaie, Y., Yomo, T.: Packet fragmentation and reassembly in molecular communication. IEEE Trans. Nanobiosci. **15**(3), 284–288 (2016)

A Diffusive Propagation Model for Molecular Communications: Analysis and Implementation in NS-3

Paul Calderon-Calderon[1], Eddy Zuniga-Gomez[1], Fabian Astudillo-Salinas[1], and Luis Tello-Oquendo[2]([✉])

[1] College of Engineering, Universidad de Cuenca, Cuenca, Ecuador
[2] College of Engineering, Universidad Nacional de Chimborazo, Riobamba, Ecuador
lptelloq@ieee.org

Abstract. In this research, the analysis and implementation of a diffusive propagation model for molecular communications are performed in NS-3. The work is based on the IEEE 1906.1-2015 standard recommendation, which seeks to create a reference framework for molecular communications. The standard provides a simulation module in NS-3, which contains only the components of the general structure of molecular communication and their interaction between them. The components mentioned are *Message Carrier, Motion, Field, Perturbation*, and *Specificity*. The transmitter uses CSK modulation. In the medium, Brownian motion (BM) with and without drift is used for the motion of the molecules, and intersymbol interference is considered. In the receiver, amplitude detection is used. The whole process is applied in four scenarios: Free BM, BM with drift, free BM bounded by the medium, and BM with drift bounded by the medium are considered. As a result, the pulse train of the mean concentration of molecules as a function of time at the receiver is obtained. In addition, the obtained results are compared with an investigation performed in N3Sim to validate the results. Finally, it is validated that the mean concentration at the receiver using the diffusive propagation model implemented complies with the mathematical model established by Fick's second law.

Keywords: Molecular communications · Diffusive propagation model · IEEE 1906.1-2015 · NS-3 · Brownian motion · CSK modulation · Intersymbolic interference · Amplitude detection

1 Introduction

The anatomy of the human body consists of tens of trillions of cells, which need to communicate with each other to respond to any stimulus produced by any organ [1]. In this context, one of the main advances in engineering applied to medicine is molecular communications (MCs). This type of communications

T. Nakano (Ed.): BICT 2021, LNICST 403, pp. 163–178, 2021.
https://doi.org/10.1007/978-3-030-92163-7_13

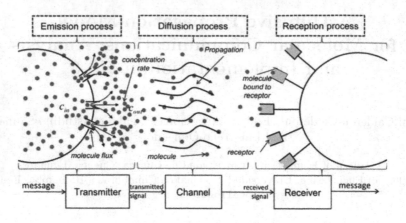

Fig. 1. Molecular communication scheme [2].

seeks to place a transmitting nanomachine and a receiving nanomachine inside the human body (see Fig. 1) [2]. The nanomachines could collect and exchange information using the same methods employed by the body at the microscopic level, such as diffusion (through molecules), motor proteins (through vesicles), or bacterial motility (through bacteria) [3].

MCs are fundamental in the near future for personalized medicine [4,5]. In this way, diseases could prevented by monitoring health in real time and highly effective treatments could carried out [6]. In turn, it will be possible to counteract a virus or bacteria before it gains strength, thanks to the sensing given by the nano-sensors.

According to [3], the research in MCs is divided into fundamental research, technology development and technology demonstration. Fundamental research (the most developed area) comprises the theoretical basis of MCs. On the other hand, technology development is the current area where researchers are focused. Technological development encompasses various activities such as physical layer signals, MAC layer protocols, chemical reactions and coding, and transmitting and receiving hardware. Finally, technology demonstration involves the design and fabrication of prototypes.

Due to the high costs involved in experimentation in MCs, research work in these sub-areas is completely theoretical and relies on simulation models for the analysis, design and optimization of mathematical or analytical models [7]. The most commonly used simulators in MCs are *NanoNs*, *MUCIN*, *N3Sim*, *dMCS*, *BINS*, *nanoNS3* and *BNsim* [3].

Currently, there is no implementation of a diffusion-based propagation model in NS-3. The closest work found in the literature is a simulation with a bacterial propagation model [8]. The main advantage of implementing an MC module in NS-3 lies in complementing the physical layer with higher layers of communication; this would allow establishing a complete simulation of MCs between two nano-nodes. Table 1 shows a comparison between the different MC simulators with their main features.

Table 1. Comparison between the different simulators of MCs [3].

Characteristics	NanoNS	nanoNS3	MUCIN	N3Sim	dMSC	BINS	BNSim
Programming language	NS-2, C++, TCL	C++	MATLAB	Java	Java	Java	Java
Parallelization	NO	NO	NO	NO	YES	NO	NO
Open source	NO	YES	YES	YES	NO	NO	YES
Propagation	Diffusion	Bacteria	Diffusion	Diffusion	Diffusion	Diffusion	Bacteria
Reception	Berg, Gillespie	Bacterial receiver	Absorption	Sampling	Absorption	Receivers	Receivers
Environment dimensions	3-D	3-D	1-D, 2-D, 3-D	2-D, 3-D	3-D	3-D	3-D
Interaction of molecules	NO		NO	YES	NO	YES	NO
Sending of consecutive symbols	NO		YES	YES	NO	NO	YES

In general, researchers do not adhere to a common framework and this hinders the integration of research. For this reason, the IEEE proposes the 1906.1-2015 standard as a recommended practice for the nanoscale and the MC framework [9].

The standard proposes a reference model for standardization of MCs to facilitate the integration in future research. As a complement, the standard provides a NS-3 module for MCs between a transmitting nano-node and a receiving nano-node. This simulation model is limited to transmitting a message, as long as the transmission rate does not exceed the channel capacity. Moreover, it works in an idealistic way, because no communication channel with losses or molecular motion is implemented. That is, it is assumed that all molecules reach the receiver and that there is no collision between them in their trajectory.

2 Design and Implementation

The simulator developed in NS-3 is based on the P1906 module. Substantial modifications were made to this module. The module initially contained only a general architecture of the MCs, consisting only by the components depicted in Fig. 2.

a) Component 0: *Message Carrier*
   The carrier contains and transports the information, in this case the molecules. An active carrier generates its own movement and a passive carrier propagates through the medium in which it is found.
b) Component 1: *Motion*
   It defines the motion capability for component 0. The motion provides the force required, regardless of direction, to transport the information efficiently through the communication channel.

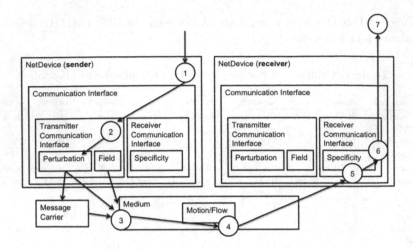

Fig. 2. Simulator architecture in NS-3 [9].

c) Component 2: *Field*
   It defines the organized movement of component 1, i.e. it gives the direction to the carrier so that it can reach the destination.
d) Component 3: *Perturbation*
   It provides the service of different carriers, as needed, to represent a signal. This is considered modulation.
e) Component 4: *Specificity*
   It defines the reception of the targeted carriers of component 3, performs carrier detection or reception when it has reached the target, improves the carrier's connectivity to its intended receiver, and helps measure the efficiency of carriers arriving at their receiver.

The implementation performed is presented in the schematic of Fig. 3.

Application mol-example: In this file the simulator application is developed. The components of the P1906 module and the NS-3 are used here. It has been modified the application implemented in the file `mol-example.cc`. The following parameters have been defined for the simulation:

- `nodeDistance` → Distance between the transmitting node and the receiving node(s).
- `nbOfMol` → Number of molecules used by the transmitter to modulate a symbol.
- `pulseInterval` → Pulse width to transmit and sense a symbol.
- `diffusionCoefficient` → Diffusion coefficient of the medium.
- `driftVelocity` → Drift velocity of the medium.
- `driftAngle` → Drift angle of the medium.
- `timeStep` → Time interval between simulation steps.
- `receptionRadio` → Reception Radio.

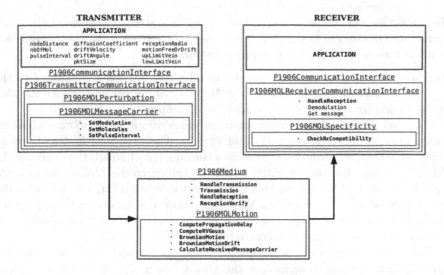

Fig. 3. Programming scheme.

- `motionFreeOrDrift` → Defines the type of diffusion: (0) an MC with free diffusion or (1) an MC with drift.
- `upLimitVein` → Upper limit of the vein.
- `lowLimitVein` → Lower limit of the vein.

The nodes and devices are created along with their related components. Then, the common component *Motion* and the *Medium* are created. For this application, a constant mobility model is considered. Next, the module's *Helpers* are used to connect the whole system. Finally, the information packets are created and the function `P1906CommunicationInterface::HandleTransmission` `(Ptr<Packet> p)` is used to send the message to the communication interface of the transmitting node.

CommunicationInterface: Implements the `HandleTransmission (Ptr <Packet> p)` function to send the packet to the transmitting interface.

Transmitter Communication Interface: It receives the message carrier properly modulated by perturbation. Then, it calls the function `P1906Medium::HandleTransmission` to send the carrier together with the *Field* component to the medium, and finally, it starts the transmission.

Component Perturbation: Within the function `CreateMessageCarrier` `(Ptr<Packet> p)` creates the carrier and performs the on-off keying (OOK) modulation to the bit packet. For the bit 1, it modulates `nbOfMol` molecules; and for the bit 0, it modulates 0 molecules.

Field Component: It receives the speed and drift angle parameters from the application. This component sends the received parameters to the *Medium* component, which in turn sends them to the *Motion* component for use in the BM with drift.

Message Carrier Component: The carrier has attributes such as the pulse interval, the number of molecules that are used for modulation, the transmission duration time and a vector representing the modulation.

Medium Class: This is the most important class. This is the channel of the MCs. As shown in Fig. 2, it contains the *Motion* component and takes data from the *Message Carrier*, *Perturbation* and *Field* components.

Four main functions are implemented in the *Medium*: `HandleTransmission`, `Transmission`, `HandleReception`, and `Verify`. The functions first and second perform the transmission of the molecules through the channel; they execute the BM of the molecules with or without drift by calling to the *Motion* component; it performs the BM for the n_0 molecules released by each bit, i.e., all molecules are released at $t_0 = 0$ and at each Δt they move one position; then, for each position traveled, it is verified whether or not each molecule has reached the receiver. The increments are generated in parallel.

The third function, `HandleReception`, is called by the function `HandleTransmission` to terminate the transmission. A call to the communication interface of the receiver is made inside `HandleReception`. With this, the `Medium` delivers the message. The fourth function, `Verify`, is an extension of the receiver that is used to verify at each time step whether or not the molecules reached the receiver.

Motion Component: This component contains the functions related to the movement of the molecules. First, there is the function `ComputePropagationDelay`, in charge of calculating the *delay*, the time when the mean concentration of molecules in the receiver is maximum. To calculate the *delay*, Fick's second law (2-D) [Eq. (1)], is derived and the result is equaled to zero. From the resulting expression the time is cleared and Eq. (2) is obtained as follows

$$C = \frac{M}{4\pi Dt} e^{\frac{-r^2}{4Dt}} \tag{1}$$

$$delay = \frac{r^2}{4D} \tag{2}$$

where, r is the distance between the transmitter and the receiver, and D is the diffusion coefficient.

Second, there is the function `ComputeRVGauss`, responsible for calculating the Gaussian random variable for the BM using a normal distribution with $\mu = 0$ and $\sigma^2 = 2\pi$ [see Eq. (3)]. Next, there are the functions `BrownianMotion` and `BrownianMotionDrift`, which perform the BM of the molecules in a medium without drift and in a medium with drift, respectively. Both functions take the random variable resulting from `ComputeRVGauss` to define the direction of motion at each step a molecule takes.

Initially, the function `BrownianMotion` calculates the random variable θ. Then, using Eqs. (4), it calculates the next position of the molecule:

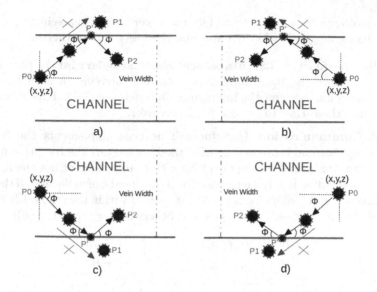

Fig. 4. Bounce of a molecule with the vein wall.

$$f_x(x) = \frac{1}{2\pi} exp\left(-\frac{x^2}{4\pi}\right) \tag{3}$$

$$\begin{aligned} x_i &= x_{i-1} + \sqrt{4D\Delta t}\cos\theta_i \\ y_i &= y_{i-1} + \sqrt{4D\Delta t}\sin\theta_i \end{aligned} \tag{4}$$

where x_i is the current position, x_{i-1} is the previous position, D is the diffusion coefficient, Δt is the time *step* of each movement and θ_i is the random variable that determines the direction of that movement.

On the other hand, the function `BrownianMotionDrift` performs exactly the same process as in the function explained above, but it is obtained from the *Field* component, the velocity, and the drift angle [see Eq. (5)]:

$$\begin{aligned} x_i &= x_{i-1} + \sqrt{4D\Delta t}\cos\theta_i + v_x\Delta t \\ y_i &= y_{i-1} + \sqrt{4D\Delta t}\sin\theta_i + v_y\Delta t \end{aligned} \tag{5}$$

where v is the drift velocity.

In collisions, elasticity is considered to be equal to 1 and friction is considered to be equal to 0. For this reason, collisions are completely elastic and the angle of reflection is the same as the angle of incidence. Furthermore, the distance traveled by the molecule is the same at each step. The proposed scheme is shown in Fig. 4. There are four possible cases:

a) The molecule goes forward and hits the upper wall of the vein.
b) The molecule goes backwards and hits the upper wall of the vein.

c) The molecule goes forward and hits the lower wall of the vein.

d) The molecule goes backwards and hits the lower wall of the vein.

Finally, the function `CalculateReceivedMessageCarrier` is part of this component, where the total number of molecules arriving in each bit is calculated, taking into account the intersymbol interference (ISI). This information is loaded into the carrier to be sent to the receiver.

Receive Communication Interface: This class implements the function `HandleReception` and is invoked by `ns3::Medium` to start the reception process. `HandleReception` uses the component *Specificity* to receive or not the message.

The demodulation is performed based on a threshold k of molecules. If the mean concentration of received molecules $c(r, t)$ at a time t within the interval T_s exceeds the threshold, the received symbol is $s = 1$, otherwise $s = 0$ [see Eq. (6)].

$$
\begin{aligned}
\text{If} \quad c(r,t) \geq k \quad &\rightarrow \quad s = 1 \\
\text{If} \quad c(r,t) < k \quad &\rightarrow \quad s = 0
\end{aligned}
\tag{6}
$$

Specificity Component: It checks if the channel capacity is greater than the transmission rate. If it is fulfilled, reception and demodulation is performed, otherwise not. According to [10], the channel capacity is given by $C_{channel} = 1/T_m$; where, T_m is the minimum pulse width for smooth transmission.

On the other hand, in [10] it is also stated that the transmission rate for diffusive molecular communications is given by $RB = 1/T_s$; where, T_s is the pulse width at which a symbol is transmitted. Then, the minimum condition that must be met for the message to be received correctly is given by $C_{channel} \geq RB$. Finally, the simulator is available for free use at https://github.com/eddyzg7/nanoDMC.

3   Results

In this section, firstly the results obtained are compared with that of the research carried out in [11] for the sake of validation. Then, four simulation scenarios are defined: free BM and BM with drift in an unconstrained medium and, free BM and BM with drift in a constrained medium. The results of the simulations has been analyzed. Within each scenario, a scheme consisting of a transmitting nanomachine, a receiving nanomachine and a medium through which the molecules will propagate is proposed. Depending on the bit to be sent, the nanotransmitter will release a certain number of molecules into the medium, so that through the BM they propagate and reach the receiver, where the information is demodulated based on a threshold.

3.1   Validation

To validate the simulator, the obtained results are compared with that of [11], which uses the N3Sim simulator. Table 2 compares the parameters used in [11] and the parameters that has been used in our simulations.

Table 2. Comparison of simulation parameters.

Parameter	Value in [11]	Value in the simulation
Time step	2 ms	2 ms
Simulation time	5 s	5 s
Transmission distance	50 μm	50 μm
Transmitter radio	5 μm	-
Receiving radius	5 μm	5 μm
Particle radius	0.2 nm	-
Number of molecules	100000 mol	100000 mol
Diffusion coefficient	$1 \frac{nm^2}{ns}$	$1 \frac{nm^2}{ns}$
Type of movement	BM free	BM free
Delimitation of veins	NO	NO

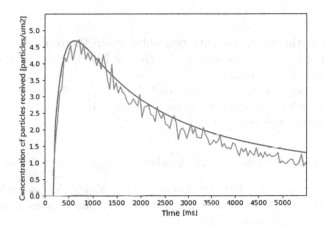

Fig. 5. Mean concentration of molecules in the receiver.

As shown in Table 2, there are parameters that are not considered in this simulator, such as the radius of the transmitter and the radius of the particle. Simulations were performed with the parameters shown in Table 2.

The result is illustrated in Fig. 5, where two curves are observed. The orange line is the mean concentration of molecules obtained by the simulator. The blue line is the analytical mean concentration of molecules established by Eq. (1).

It is observed that both curves are similar. The amplitude value is adjusted during the whole simulation time, and the maximum mean concentration in both cases is equal. Furthermore, the correlation between both curves has been calculated to be 88%. The implementation performed considers that the molecules are released simultaneously; this generates a pulse similar to a *Dirac delta* function $\delta(k)$, with small pulse width.

On the other hand, Fig. 6 illustrates the results obtained by [11]. The blue line represents the analytical result obtained with Fick's second law [Eq. (1)].

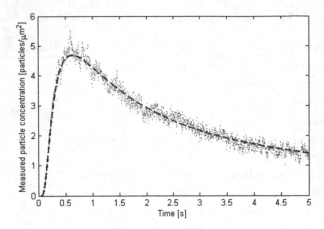

Fig. 6. Mean concentration of molecules in the receiver obtained by [11].

Each red dot is the mean concentration obtained in the receiver per unit time. If the results of Fig. 5 are compared with that of Fig. 6, it can be observed that the results are quite similar.

From this comparison, it is concluded that the results are validated, since they coincide with the mathematical model defined by Fick's Law, and with the results obtained in the research [11].

3.2   Simulation Scenarios and Analysis

Table 3 indicates the simulation parameters. The simulation is performed for a free and drifting BM. Note that enough number of simulations were conducted to validate the results statistically.

The variable `motionFreeOrDrift`, varies between the values `BROWNIANMOTION` and `BROWNIANMOTIONDRIFT`, depending on whether it is simulated in a free or drifting BM, respectively. The variables `upLimitVein` and `lowLimitVein` are used only in a bounded medium.

a) Free Brownian motion

As an example, Fig. 7 illustrates the movement of one molecule that is emitted by the transmitter and reaches the receiver. The distance between the transmitter and receiver is 26 μm. The black and green circles represent the transmitter and receiver nanomachines, respectively. Due to the free motion, the molecule moves randomly over the medium, i.e., it has no direction. For this reason, the molecule is delayed in reaching the receiver, as it makes a few turns before entering the receiving area.

Each molecule released by the transmitter has a different behavior. A set of molecules is needed to transmit a bit; this set of molecules is know as a pulse. Figure 8 depicts the train of pulses arriving at the receiver. These pulses represent the bit stream sent. The receiver demodulates the information based

Table 3. Simulation parameters.

Parameter	Variable name	Value	Unit
Distance between Tx and Rx	nodeDistance	26	µm
Number of molecules	nbOfMol	10000	mol
Pulse width	pulseInterval	1050	ms
Diffusion coefficient	diffusionCoefficient	1	$\frac{nm^2}{ns}$
Drift velocity	driftVelocity	100	$\frac{\mu m}{s}$
Drift angle	driftAngle	0	rad
Number of intervals	timeSlot	260	slots
Receiving radius	receptionRadio	1	µm
Type of movement	motionFreeOrDrift	BROWNIANMOTION BROWNIANMOTIONDRIFT	
Upper limit of the vein	upLimitVein	50	µm
Inferior limit of the vein	lowLimitVein	−50	µm
Bitstream size	pktSize	1	Byte
Message to be transmitted	buffer	11011001	Bits

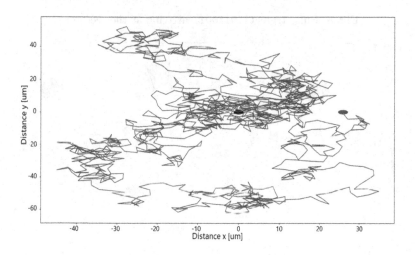

Fig. 7. Free BM of a molecule that reaches the receiver.

on the threshold. The molecules that arrive for the bits with a value of 0 are the result of the ISI, but since the amplitude is not large enough, they can be considered as white noise.

b) Brownian motion with drift
The movement of one molecule is represented in Fig. 9 in a similar way as free BM was represented, using the same parameters except for the drift. At this scenario, the molecule is being influenced by the blood flow which gives it direction, and thus arrives faster at the receiver.

Figure 10 illustrates the train of pulses received for the complete transmission of the message. The pulses are clearly defined for a bit 1 and a bit 0 with the drift. In contrast to the free BM (see Fig. 8), the mean concentration curve is

P. Calderon-Calderon et al.

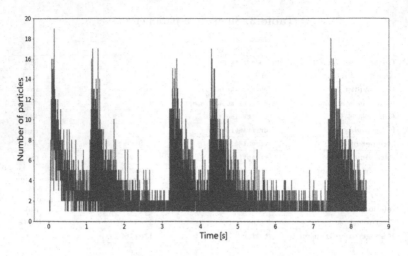

Fig. 8. Pulse train (11011001) arriving at the receiver.

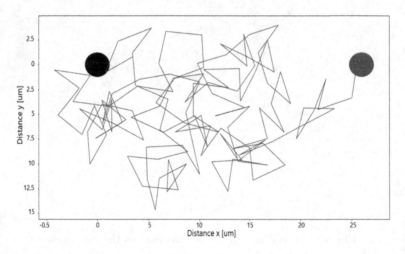

Fig. 9. BM with derivatization of a molecule that reaches the receiver.

more sharply defined due to the decrease in random noise in the drifted BM. It is shown that a more significant number of molecules arrive.

c) Free brownian motion bounded by the width of the medium

In addition to the implementation of the diffusive propagation model in NS-3, a medium bounded by blood vessel walls has been proposed. This medium limits the movement of the molecules. When the medium is large enough with respect to the size of the nanomachines, it can be considered as an infinite plane; however, when the sizes are comparable, the medium limits the motion of the molecules. The implemented model is detailed in Sect. 2. According to [12], the diameter of a small artery called an arteriole is 100 μm. For this reason, a value

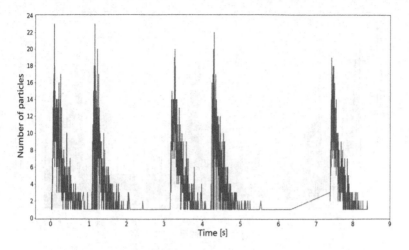

Fig. 10. Pulse train (11011001) arriving at the receiver.

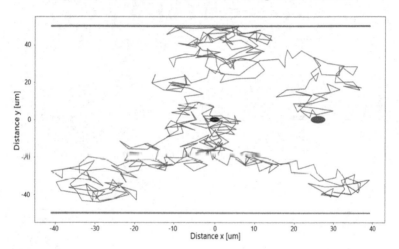

Fig. 11. Free BM of a molecule reaching the receiver, bounded by the walls of the arterioles.

of 50 μm is placed at each boundary. In Fig. 11 the result of the free BM bounded by the walls of the arterioles is visualized.

The behavior of the molecule remains chaotic; however, by limiting the space through which it moves, the possibility of reaching the receiver increases. When the molecule reaches the arteriole wall, it collides elastically and takes a new direction until it stabilizes. The effect of the walls will be reflected in the increase of the mean concentration of molecules in the receiver.

Figure 12 depicts the number of molecules in the receiver over the time, for the transmission of an 8-bit stream. It can be seen that there is more interference between bits and the pulse amplitude has increased.

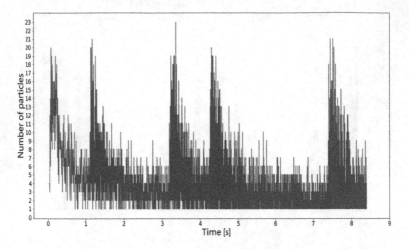

Fig. 12. Pulse train (11011001) at the receiver in a bounded medium for free BM.

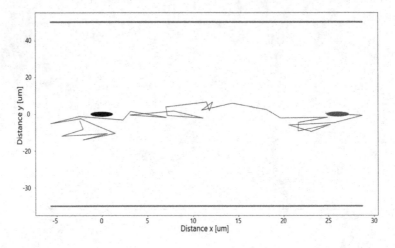

Fig. 13. BM derived from a molecule that reaches the receiver in a medium bounded by the walls of the arterioles.

d) Brownian motion with drift bounded by the width of the medium

For a BM with drift, Fig. 13 visualizes the simulation result of the movement of a molecule arriving at the receiver. In this case, the molecule does not touch the walls of the arterioles and reaches the receiver without so much travel in the medium. However, the behavior of each molecule is different. Therefore, other molecules will not reach the receiver as fast.

For a complete transmission of a bit stream, one has the pulse train shown in Fig. 14. The results show well-defined pulses with minimal interference between bits; this result is characteristic of the BM with drift. In this case, the walls of the arterioles increase the amplitude of all received pulses.

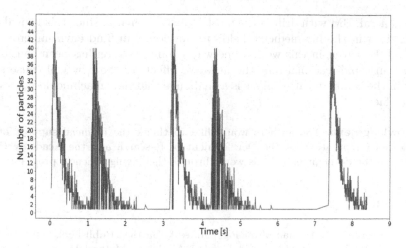

Fig. 14. Pulse train (`11011001`) at the receiver for a BM with drift in a bounded medium.

4 Conclusions

A diffusive propagation model for MCs has been implemented in NS-3. Seven simulators have been found for MCs, but none use the IEEE 1906.1-2015 recommendation. In this work, the recommended standardization process is followed.

The results obtained from the simulations have been shown in Sect. 3. In each scenario, the motion of a particle and the received pulse train of the whole message are analyzed. The results show that the BM with drift is characterized by less chaotic motion and high particle concentration at the receiver due to steering by the medium flow. The pulses at the receiver are better defined and have little interference from molecules. On the other hand, the free BM is characterized by higher randomness in the movements, fewer molecules in the receiver, and higher noise due to delayed molecules.

In the third and fourth scenarios analyzed, the medium is bounded by the walls of an arteriole. For a bounded free BM, the results show that the number of molecules in the receiver increases considerably, as does the noise. On the other hand, for a BM with bounded drift, the mean concentration increases twice as much as in the previous case. The pulses are more defined, and there is less noise due to loose molecules.

In this work, inter-symbol interference was implemented. In this case, ISI is conceived as the effect produced by the molecules of a bit that do not reach the receiver and remain free in the medium. The results show that the interference is higher for the bounded or unbounded free BM scenario. Since a flow does not orient the molecules, they remain in the medium and arrive bit by bit in a non-corresponding bit time. To decrease this effect, one can use different modulations or consider a lifetime for the molecules. However, in most cases, it implies a more complex transmitter design, contrary to what is sought in MCs.

In general, BM with drift is the most realistic scenario, since it is logical that applications in the bio-medicine field are developed in fluid environments such as blood. However, in this work a one-way communication has been considered because in a bidirectional one, the negative effects of the flow will cause some issues in the communication. This is a challenge that nanomachine designers will have to face.

Acknowledgment. The authors would like to thank the financial support of the Ecuadorian Corporation for the Development of Research and the Academy (RED CEDIA) in the development of this work through the Divulga Ciencia program.

References

1. Silverthorn, D.U.: Human physiology. Jones & Bartlett Publishers, Oxford (2015)
2. Suzuki, J., Nakano, T., Moore, M.J. (eds.): Modeling, Methodologies and Tools for Molecular and Nano-scale Communications. MOST, vol. 9. Springer, Cham (2017). https://doi.org/10.1007/978-3-319-50688-3
3. Farsad, N., Yilmaz, H.B., Eckford, A., Chae, C., Guo, W.: A comprehensive survey of recent advancements in molecular communication. IEEE Commun. Surv. Tutorials **18**(3), 1887–1919 (2016)
4. Murga, C.: Mecanismos moleculares de comunicación entre células: claves para el diseño de nuevos fármacos y la medicina personalizada. In: SEBBM. Dpto. de Biología Molecular de la Universidad Autónoma de Madrid y Centro de Biología Molecular "Severo Ochoa", no. 1 (2015)
5. Cevallos, Y., Tello-Oquendo, L., Inca, D., Ghose, D., Shirazi, A.Z., Gomez, G.A.: Health applications based on molecular communications: a brief review. In: 2019 IEEE International Conference on E-health Networking, Application and Services (HealthCom), pp. 1–6. IEEE, (2019)
6. Yang, K., et al.: A comprehensive survey on hybrid communication in context of molecular communication and terahertz communication for body-centric nanonetworks. IEEE Trans. Molecular, Biological Multi-Scale Commun. **6**(2), 2–3 (2020)
7. Nakano, T., Eckford, A.W., Haraguchi, T.: Molecular Communication. Cambridge University Press, Cambridge vol. 1, pp. 71–96 (2013)
8. Jian, Y., et al.: nanons3: A network simulator for bacterial nanonetworks based on molecular communication, Nano Communication Networks, vol. 12, pp. 1–11 (2017). https://www.sciencedirect.com/science/article/pii/S1878778916300941
9. Nanoscale, N., Molecular Communications, W.G.: IEEE recommended practice for nanoscale and molecular communication framework. IEEE Std 1906.1-2015, pp. 1–64 (2016)
10. Llatser, I., Cabellos-Aparicio, A., Pierobon, M., Alarcon, E.: Detection techniques for diffusion-based molecular communication. IEEE J. Sel. Areas Commun. **31**(12), 726–734 (2013)
11. Llatser, I., et al.: Exploring the physical channel of diffusion-based molecular communication by simulation. In: 2011 IEEE Global Telecommunications Conference - GLOBECOM 2011, pp. 1–5 (2011)
12. Informed. Sistema cardiovascular, Red de Salud de Cuba. Centro Nacional de Información de Ciencias Médicas, pp. 7–13, (2021). http://www.sld.cu/galerias/pdf/sitios/histologia/sistema_cardiovascular.pdf

Electromagnetism-Enabled Transmitter of Molecular Communications Using Ca²⁺ Signals

Peng He[1,2,3](✉) and Donglai Tang[4]

[1] School of Communication and Information Engineering, Chongqing University of Posts and Telecommunications, Chongqing, China
[2] Key Laboratory of Optical Communication and Networks in Chongqing, Chongqing, China
[3] Key Laboratory of Ubiquitous Sensing and Networking in Chongqing, Chongqing, China
[4] Aostar Information Technologies Co., Ltd., Beijing, China
tangdonglai@sgitg.sgcc.com.cn

Abstract. Molecular Communications provides a promising solution to achieve precise control and process of bio-things in applications of Healthcare-IoT. In this paper, we investigates the mechanism of electromagnetism-induced molecular communications (EMC) among non-excitable biological cell networks. We choose calcium signals as the physical information carrier to study the paradigm of EMC. Firstly, an electromagnetism-potential coupling model is established to study the electric-magnetic induction behaviour of cellular membrane potential. Then, an Ca^{2+} oscillation model is established to study the relation between membrane potential and Ca^{2+} signals. Further, we validate the waveform patterning of calcium signaling by applying various intensities and frequencies of electromagnetism. This paper shows the relations between electromagnetism stimuli and calcium oscillation through mathematical modeling and numerical experiments. We find that there exists a resonance behavior between electromagnetism and calcium signals, namely calcium signals oscillate via a similar frequency with the electromagnetism. This paper reveals that molecular communication can be effectively induced by traditional electromagnetic signals.

Keywords: Molecular communications · Electromagnetism-enabled · Calcium signals · Transmitter design

Supported by the National Natural Science Foundation of China (Grant No. 61901070, 61871062, 61771082, 61801065), partially supported by the Science and Technology Research Program of Chongqing Municipal Education Commission (Grant No. KJQN201900611, KJQN201900604), and partially supported by Program for Innovation Team Building at Institutions of Higher Education in Chongqing (Grant No. CXTDX201601020).

T. Nakano (Ed.): BICT 2021, LNICST 403, pp. 179–189, 2021.
https://doi.org/10.1007/978-3-030-92163-7_14

1   Introduction

Progress of precise healthcare applications requires nano-scale process technologies of bio-things in intra-body environment, where most traditional IoT devices are not tiny and bio-compatible enough for intra-body applications [1]. Molecular communications (MC) are proposed as a new communication paradigms which enables multiple of nano-machines to cooperate for various nano-scale tasks [2]. In a MC system, nano-machines act as different roles including transceiver and relays to process information of nano-scale networks.

In this work, we choose Ca^{2+} signals as the information carrier for MC communication system due to follow two considerations. First, Ca^{2+} signals is one of the most important physiological signals in vivo. Ca^{2+} signals is the well-known second messenger, which plays critical roles in many metabolism processes and is closely associated with many diseases. Thereby Ca^{2+} signals are worthy to be investigated in the research scope of biology and medicines. Second, Ca^{2+} signals appear as some patterns of waveforms, that is similar to the electromagnetic wave. In the ideal environment, Ca^{2+} signals even propagate as the periodic waves of impulses, the amplitudes and frequency of impulses can be regulated via some bio-engineering methods.

A simplest MC system can be divided into a transmitter, a receiver and the communication channel according to the physical layer concept. Plenty of literatures have addressed on the work of receiver technologies, which aims to extract the information from the absorbed molecules around the receivers. Typical cases are like signal detection [3], decoding or demodulation [4], ISI-elimination [5], etc. Research on MC channel enables nano-machines adapts the unique characters of the special channels. Typical cases are channel coding [6], channel modelling [7], channel synchronization [8], etc. Transmitter is another important unit in communication system. However, limited literatures try to study MC from the aspect of transmitter. In [9], a multiple transmitter local drug delivery system associated with encapsulated drug transmitters was investigated. In [10], the authors discussed issues concerned with transmission rate control in molecular communication, an emerging communication paradigm for bio-nanomachines in an aqueous environment. In [11], the authors modeled the molecule emission process more accurately using rectangular and exponential transmit signals and derived closed form expressions for the number of molecules that are absorbed by the receiver in a diffusion-Based MC system.

To design the transmitter, an unsolved challenge is how to trigger and control nano-machines directly in bio-systems. Embedded nano-machines placed in various organs, tissues and cells, which need programmable instructions. One major class of existing solutions are enabling nano-machine with intelligent abilities via some bio-engineering methods. For example, DNA robots equipped with recombinant nucleotides are capable of storing, moving and reacting in the biological networks [12]. Bacteria can also be modified using filled nucleotides to express various functions of nano-machines. Another way to trigger biological signals can be from external space. Plenty of experiment results reveal that some physical or chemical stimuli can effectively induce various biological signals and promote metabolism of bio-systems [13].

In this work, we consider a non-excitable epidermis cell as the transmitter in MC system. We propose a controllable designed transmitter for MC system. With the inspiration of external electromagnetism, the transmitter cell is able to generate various patterns of waveforms in the communication process in the design. By adjusting the parameters of electromagnetism, we can adjust the amplitude and frequency of the waveforms, in order to express the various bits in the coding. We establish a model of Ca^{2+} signal generation, which is divided into electromagnetism-potential coupling phase and Potential-induced Ca^{2+} signal phase. In the first phase, the electromagnetism act on the cell membrane and elevate the its potential. In the second phase, the elevated membrane potential promote the release of Ca^{2+} ions from organelles to generate MC signals.

The rest of this paper is organized as follows. Section 2 introduces the transmitter design of MC. Section 3 shows numerical results to demonstrate the capability and performance of the designed transmitters. Section 4 gives a summary of this work to conclude the paper.

2   System Overview

The proposed MC system is presented in Fig. 1, which is composed of an electromagnetic device and biological epidermis cells belonging to human-beings. We assume that the electromagnetic device is regarded as the controller, which is capable to actuate the transmitter by emitting low-frequency electromagnetic waves. The low-frequency electromagnetic waves are regarded as the controlling massages, which effects on biological cells and lead a series of bio-chemical reactions epidermis cells. We choose sine save to contain the controlling massages due to its university in application of wireless communications. Epidermis cells are regarded as the transmitter, where Ca^{2+} concentration of epidermis cells oscillate as various patterns of waveforms to trigger MC process. In this work, we focus on the transmitter design in one single cell, namely, we ignore the interactions between different connected cells.

3   Electromagnetism-Enabled Transmitter Design

The proposed transmitter includes two phases, which are electromagnetism-potential coupling phase and potential-induced Ca^{2+} signaling phase.

3.1   Electromagnetism-Potential Coupling Phase

Let $I = E_m sin(2\pi ft)$ denotes the alternating electromagnetism as the system input with electromagnetism intensity E_m and alternating frequency f. The potential of the cell membrane is raised due the movement of electric ions, which is promoted by three different classes of forces, expressed by,

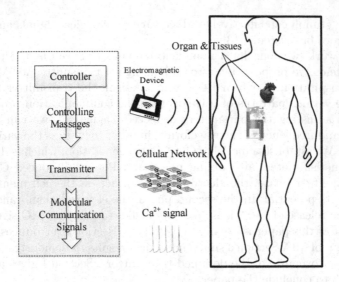

Fig. 1. Proposed system model

$$m\frac{d^2l}{dt^2} - F_3 + F_1 = F_2 \tag{1}$$

where l is the moving distance of electric ion compared to the initial position during the moving process.

The first class of force is restoring force resulted by electrochemical gradient,

$$F_1 = -m\omega^2 l \tag{2}$$

where m is the quality of the electric ion, and $\omega = 2\pi f_s$ is the self-sustained oscillation angular frequency. Accordingly, f_s is the self-sustained oscillation frequency.

The second class of force is the external force brought by the electromagnetism, expressed by,

$$F_2 = E_m z q_e sin(2\pi ft) \tag{3}$$

where z is the valence state of the electric ion. $q_e = 1.6 * 10^{-19}$ C is the unit electric charge.

The third class of force is decaying force, expressed by,

$$F_3 = -\lambda\frac{dl}{dt} \tag{4}$$

where λ is the attenuation coefficient. The approximated solution of Eq. (1) is given by,

$$l = -\frac{E_m z q_e}{2\pi\lambda v}\cos(2\pi vt) \tag{5}$$

In this work, we consider the effect of electromagnetism on SOC channel on the cellular membrane, resulted by a force [],

$$F = \frac{1}{4\pi\varepsilon\varepsilon_0} \cdot \frac{q \cdot zq_e}{r^2} \tag{6}$$

where ε_0 is the vacuum dielectric constant, ε is the relative dielectric constant, $q = 1.6 * q_e$ is the charge of SOC channel. r is the distance between the free charge and SOC channel. Cellular membrane potential is related to the membrane thickness s, the effect force F and the charge q, expressed by,

$$V_m = F \cdot \frac{s}{q} \tag{7}$$

Through a differential operation, we have,

$$\partial V_m = \partial F \cdot \frac{s}{q} \tag{8}$$

Take the expression of Eq. (6) into (8), we have,

$$\partial V_m = \frac{1}{2\pi\varepsilon\varepsilon_0} \cdot \frac{q \cdot zq_e}{r^3} \partial r \tag{9}$$

Assume that position of the electrical ion is the origin, then $\partial x = \partial r$ holds. By integrating Eq. (5) and (8) into (9), the variation of membrane potential is give by,

$$\partial V_m = \frac{1}{2\pi\varepsilon\varepsilon_0} \cdot \frac{q \cdot zq_e}{r^3} \cdot \frac{s}{q} \cdot \frac{E_0 zq_e}{\lambda} \sin(2\pi vt)\partial t \tag{10}$$

3.2   Potential-Induced Ca^{2+} Signaling Phase

A widely-accepted model [16] of calcium dynamic is adopted to describe the intracellular oscillation process, which is shown in Fig. 2. For cell i in the network, two variables x_i and y_i are respectively used to describe the Ca^{2+} concentrations in the cytoplasm and internal store. The Ca^{2+} dynamics is described by the set of equations below.

$$\frac{dx}{dt} = P_1 - P_2 + P_3 + a_1 y - a_2 x + I_G, \tag{11}$$

$$\frac{dy}{dt} = P_2 - P_3 - a_1 y, \tag{12}$$

where

$$Q_1 = b_1 + b_2(t) \tag{13}$$

$$P_2 = \frac{b_3 \cdot x^2}{b_4{}^2 + x^2}, \tag{14}$$

$$P_3 = \frac{b_5 \cdot x^4 y^4}{(b_6{}^4 + x^4)(b_7{}^2 + y^4)}. \tag{15}$$

P_1 denotes the Ca^{2+} increase from the extracellular space via different classes of channels. P_2 and P_3 determine the Ca^{2+} exchange between the cytoplasm and internal stores due to the regulation of channel permeability. In P_1, b_1 denotes non-electromagnetism-induced Ca^{2+} increase due to the constant pumping of Ca^{2+} influx. $b_2(t)$ denotes Ca^{2+} increase due to the promotion of electromagnetism. In this model, the expression of $b_2(t)$ is related with change of membrane potential V_m,

$$C_m \frac{dV_m}{dt} = b_2(t) \tag{16}$$

where C_m is the capacitance of the cell membrane.

4   Performance Evaluation

In this section, we present the patterning behaviours of Ca^{2+} signals induced by the transmitter. The target of the simulation is to validate the availability and performance of the proposed transmitter.

4.1   Simulation Configuration

The parameter configures include two parts: the first part is parameters of electromagnetism-potential coupling phase, which is listed in Table 1 [14,15]; the second part is parameters of potential-induced Ca^{2+} signaling phase, listed in Table 2 [16,17].

Table 1. Electromagnetism-potential coupling parameters

Parameter	Value
q_e	10^{-19} C
q	$1.7 * q_e$ C
z	1
ε	4
ε_0	$8.854 * 10^{-12}$ $N^{-1}m^{-2}C^2$
r	10^{-9} m
s	10^{-8} m
λ	$6.4 * 10^{-12}$ kg/s

Table 2. Ca^{2+} signaling parameters

Parameter	Value
a_1	$1\ \mathrm{s}^{-1}$
a_2	$6\ \mathrm{s}^{-1}$
b_1	$1.3\ \mu\mathrm{ms}^{-1}$
b_3	$65\ \mu\mathrm{ms}^{-1}$
b_4	$1\ \mu\mathrm{m}$
b_5	$500\ \mu\mathrm{ms}^{-1}$
b_6	$0.9\ \mu\mathrm{m}$
b_7	$2.0\ \mu\mathrm{m}$

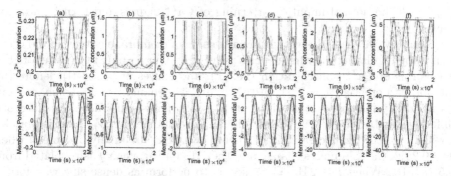

Fig. 2. Waveform patternings of membrane potential (V_m) and Ca^{2+} signals (x) under different E_m, $f = 0.2\,\mathrm{Hz}$, (a) (g) $E_m = 0.0005\,\mathrm{mV/m}$, (b) (h) $E_m = 0.002\,\mathrm{mV/m}$, (c) (i) $E_m = 0.005\,\mathrm{mV/m}$, (d) (j) $E_m = 0.01\,\mathrm{mV/m}$, (e) (k) $E_m = 0.05\,\mathrm{mV/m}$, (f) (l) $E_m = 0.01\,\mathrm{mV/m}$.

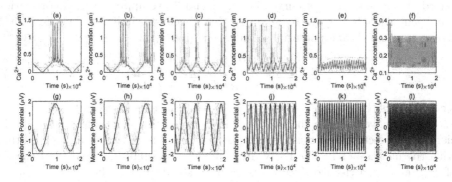

Fig. 3. Waveform patternings of membrane potential (V_m) and Ca^{2+} signals (x) under different f, $E_m = 0.005\,\mathrm{mV/m}$, (a) (g) $f = 0.08\,\mathrm{Hz}$, (b) (h) $f = 0.1\,\mathrm{Hz}$, (c) (i) $f = 0.2\,\mathrm{Hz}$, (d) (j) $f = 0.5\,\mathrm{Hz}$, (e) (k) $f = 1\,\mathrm{Hz}$, (f) (l) $f = 2\,\mathrm{Hz}$.

4.2 Waveform Patterning Presentation

In Fig. 2, we check the waveform patterning of membrane potential and Ca^{2+} signaling by regulating E_m. Curves of (a) to (f) denote the waveforms of Ca^{2+} concentration (i.e., x), and curves of (g) to (i) denote the membrane potential (i.e., V_m). We can see that V_m changes as a sine shape, that is also presented in Eq. (10). Increase of E_m amplifies the amplitude of V_m from 0.2 to 40 μV, while its oscillation frequency keep the same with f (4 periods in 20 s). By observing the Ca^{2+} concentration x, we see that x performs as sine wave when E_m is small ($E_m = 0.0005\,\mathrm{mV/m}$), then becomes the impulse arrays when E_m is a little larger ($E_m = 0.002, 0.005\,\mathrm{mV/m}$), and becomes approximated sine waveforms when E_m is large enough $E_m = 0.01, 0.05, 0.1\,\mathrm{mV/m}$. According to existing knowledge, Ca^{2+} signals perform as impulse arrays when cells function normally. So we have the following findings: 1) Ca^{2+} signals perform as normal impulse arrays just when E_m locates in a suitable area; 2) High intensities of E_m break the normal impulse patterning of Ca^{2+} signals and turn it into sine waves.

Similarly, we check the waveform patterning of membrane potential and Ca^{2+} signals by regulating their frequencies f in Fig. 3. Curves of (a) to (f) denote the waveforms of Ca^{2+} concentration (i.e., x), and curves of (g) to (i) denote the membrane potential (i.e., V_m). In this figure, V_m changes as a sine shape, and its frequency increases under different f. When f is small, one E_m peak evoke different number of Ca^{2+} impulses. The impulse number per V_m peak is initially 4 when $f = 0.08\,\mathrm{Hz}$, decrease into 3 when $f = 0.1\,\mathrm{Hz}$, and finally becomes 1 when $f = 0.2\,\mathrm{Hz}$. Ca^{2+} impulses change and then disappear when f is regulated from 0.5 to 1 Hz. When $f = 2\,\mathrm{Hz}$, Ca^{2+} seems to perform as dense sine waveforms. We find that slow oscillation of electromagnetism evoke discrete Ca^{2+} impulses, and fast oscillation of electromagnetism turn Ca^{2+} signals into sine waves.

4.3 Waveform Similarity Analysis

Based on the similarity of curves in Fig. 2 and 3, we apply a signal correlation coefficient (denoted by ρ) to quantify the waveform similarity between Ca^{2+} signals and membrane potential signals. ρ is given by,

$$\rho = \frac{E((x - E(x))(V_m - E(V_m)))}{\sqrt{E((x - E(x))^2)E((V_m - E(V_m))^2)}} \tag{17}$$

where $E(.)$ is the average function. Obviously, ρ locates in area of $[-1, 1]$;

In Fig. 4, we validate the relationship between the signal correlation coefficient ρ and electromagnetism intensity E_m under different f. It can be seen that with the increase of E_m, ρ firstly drop quickly from 0.9 during a small change of E_m, and then rise slowly until recover to steady level. There exists a minimum point for ρ by altering E_m. Also, various f make difference for ρ. when $f = 0.1\,\mathrm{Hz}$, ρ could reach to nearly 1 if E_m is 0.02 mV/m, while ρ could just reach to nearly 0.7 when $f = 1\,\mathrm{Hz}$. The minimum point of ρ also increases with increase of f.

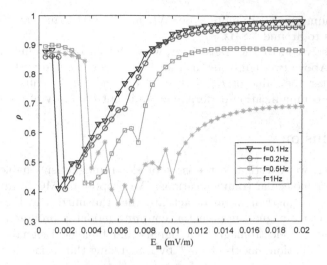

Fig. 4. Signal correlation coefficient ρ with variation of E_m under different f.

In Fig. 5, we validate the relationship between the Signal correlation coefficient ρ with variation of f under different E_m. It can be seen that with the increase of f, ρ firstly drops and then rise. There exists a minimum point for ρ by altering f. Also, various E_m make difference for ρ. when $E_m = 0.0005\,\mathrm{mV/m}$, the minimum point is around from 0.4 to 0.6, and the rise of ρ is very slow around 0. when $E_m = 0.005\,\mathrm{mV/m}$, the minimum point of f is located in $1.4\,\mathrm{Hz}$, and ρ rise quickly then fall again. Similar case holds when $E_m = 0.008\,\mathrm{mV/m}$,

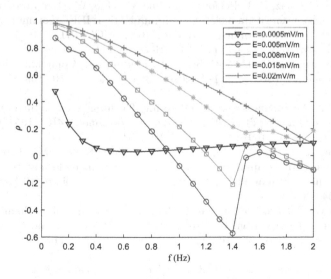

Fig. 5. Signal correlation coefficient ρ with variation of f under different E_m.

their minimum points are quite close. When $E_m = 0.015\,\mathrm{mV/m}$, the minimum point moves to around $1.8\,\mathrm{Hz}$.

To precisely control the transmitter, we wish to avoid the minimal point and increase ρ. Above two figures tell us that there are some bad performance cases when we alter the value of E_m and f. The trends of the curves guide us to increase the communication performance and controllability of the transmitter.

5   Conclusion

In this paper, we investigated the issue of electromagnetism-enabled transmitter design for molecular communications. We proposed a MC framework which utilize electromagnetism waves to actuate the transmitter. In the transmitter design, an electromagnetism-potential coupling model is proposed to establish a bridge between electromagnetism signals and biological signals. We further extend the oscillation models of Ca^{2+} by integrating the variation of membrane potential.

This paper shows that electromagnetism-enabled transmitter of molecular communications are feasible through mathematical modeling and numerical results. Future work includes the extension of proposed method from single cell to multiple cells, and experiment validation of electromagnetism-induced molecular communications.

References

1. Akyildiz, I.F., Brunetti, F., Blazquez, C.: Nanonetworks: a new communication paradigm. Comput. Netw. **52**(12), 2260–2279 (2008)
2. Farsad, N., Yilmaz, H.B., Eckford, A., Chae, C., Guo, W.: A comprehensive survey of recent advancements in molecular communication. IEEE Commun. Surv. Tutor. **18**(3), 1887–1919 (2016). https://doi.org/10.1109/COMST.2016.2527741
3. Aslan, E., Pekergin, F., Çelebi, M.E.: Receiver detection methods on molecular communications. In: 2020 28th Signal Processing and Communications Applications Conference (SIU), pp. 1–4 (2020). https://doi.org/10.1109/SIU49456.2020.9302509
4. Chou, C.T.: Maximum a-posteriori decoding for diffusion-based molecular communication using analog filters. IEEE Trans. Nanotechnol. **14**(6), 1054–1067 (2015). https://doi.org/10.1109/TNANO.2015.2469301
5. Mosayebi, R., Gohari, A., Mirmohseni, M., Nasiri-Kenari, M.: Type-based sign modulation and its application for ISI mitigation in molecular communication. IEEE Trans. Commun. **66**(1), 180–193 (2018). https://doi.org/10.1109/TCOMM.2017.2754492
6. Kışlal, A.O., Pusane, A.E., Tuğcu, T.: A comparative analysis of channel coding for molecular communication. In: 2018 26th Signal Processing and Communications Applications Conference (SIU), pp. 1–4 (2018). https://doi.org/10.1109/SIU.2018.8404368
7. Nakano, T., Okaie, Y., Liu, J.: Channel model and capacity analysis of molecular communication with Brownian motion. IEEE Commun. Lett. **16**(6), 797–800 (2012). https://doi.org/10.1109/LCOMM.2012.042312.120359

8. Luo, Z., Lin, L., Guo, W., Wang, S., Liu, F., Yan, H.: One symbol blind synchronization in SIMO molecular communication systems. IEEE Wirel. Commun. Lett. **7**(4), 530–533 (2018). https://doi.org/10.1109/LWC.2018.2793197
9. Salehi, S., Moayedian, N.S., Javanmard, S.H., Alarcón, E.: Lifetime improvement of a multiple transmitter local drug delivery system based on diffusive molecular communication. IEEE Trans. Nanobiosci. **17**(3), 352–360 (2018). https://doi.org/10.1109/TNB.2018.2850054
10. Nakano, T., Okaie, Y., Vasilakos, A.V.: Transmission rate control for molecular communication among biological nanomachines. IEEE J. Sel. Areas Commun. **31**(12), 835–846 (2013)
11. Dhayabaran, B., Raja, G.T., Magarini, M., Yilmaz, H.B.: Transmit signal shaping for molecular communication. IEEE Wirel. Commun. Lett. **10**(7), 1459–1463 (2021). https://doi.org/10.1109/LWC.2021.3069875
12. Morishima, K., Fukuda, T., Arai, F., Matsuura, H., Yoshikawa, K.: Noncontact transportation of DNA molecule by dielectrophoretic force for micro DNA flow system. In: Proceedings of IEEE International Conference on Robotics and Automation, vol. 3, pp. 2214–2219 (1996). https://doi.org/10.1109/ROBOT.1996.506493
13. Zhang, T., et al.: Molecular association between diabetes-specific local gene network and nutrient metabolism modules. In: 2010 4th International Conference on Bioinformatics and Biomedical Engineering, pp. 1–5 (2010). https://doi.org/10.1109/ICBBE.2010.5517883
14. Panagopoulos, D.J., Messini, N., Karabarbounis, A., Philippetis, A.L., Margaritis, L.H.: A mechanism for action of oscillating electric fields on cells. Biochem. Biophys. Res. Commun. **272**(3), 634–640 (2000)
15. Ming-Yan, L., Kun, S., Xu, Z., Imshik, L.: Mechanism for alternating electric fields induced-effects on cytosolic calcium. Chin. Phys. Lett. **26**(3), 017102 (2000)
16. Goldbeter, A., Dupont, G., Berridge, M.J.: Minimal model for signal-induced Ca^{2+} oscillations and for their frequency encoding through protein phosphorylation. Proc. Natl. Acad. Sci. U.S.A. **87**(4), 1461–1465 (1990)
17. Kepseu, W.D., Woafo, P.: Intercellular waves propagation in an array of cells coupled through paracrine signaling: a computer simulation study. Phys. Rev. **73**, 041912 (2014)

Research Challenges on Molecular Communication-Based Internal Interfaces for IoBNT Systems

Yutaka Okaie[✉]

Graduate School of Engineering, Osaka City University, Osaka, Japan
okaie@osaka-cu.ac.jp

Abstract. In this position paper, we describe research challenges on the Internet of Bio-Nano-Things (IoBNT), an emerging paradigm that integrates molecular communication and traditional electromagnetic communication systems into a single communication system. IoBNT systems are expected to bring innovation to conventional healthcare methods. This paper introduces new interfaces, termed internal interfaces to facilitate two-way communication between the edge interfaces and the target molecular communication systems. We assume that the target molecular communication systems are deployed deep inside the body to obtain physiological data, therefore the internal interfaces that make communication between them are essential. This paper describes research challenges associated with internal interfaces.

Keywords: Internet of Bio-Nano-Things · Molecular communication · Medical applications

1 Introduction

The advancement in nanotechnology enables us to fabricate minuscule sensors, with which we can obtain physiological data that are available deep inside in the body, but that is not well-explored due to the limitation of technologies so far. By making full use of the data obtained by bionanosensors for the benefit of human healthcare, future medical systems are envisioned to be integrated into existing computer communication systems.

A new paradigm of the Internet of Bio-Nano-Things (IoBNT) has emerged from the aforementioned background. IoBNT is a novel concept of using bionanosensors to obtain information at the molecular and cellular levels in living organisms, and sharing and controlling it via the Internet [1–3]. If IoBNT is realized, it will be possible to examine and treat patients in remote locations, and expert systems that can diagnose and treat patients without medical experts may be developed to become the medical infrastructure of society.

© ICST Institute for Computer Sciences, Social Informatics and Telecommunications Engineering 2021
Published by Springer Nature Switzerland AG 2021. All Rights Reserved
T. Nakano (Ed.): BICT 2021, LNICST 403, pp. 190–195, 2021.
https://doi.org/10.1007/978-3-030-92163-7_15

The interface bridging between molecular communication (MC) and conventional information and communication technology (ICT) systems is an essential component for realizing IoBNT. It would facilitate an independent development of both MC and ICT systems to realize as a generic interface as possible. Researchers in the field of MC have been focusing on the bio-nanomachine to bio-nanomachine interface (BNI) since its founding in 2005 as a new communication paradigm for nano-scale devices. There are in large two types of interfaces: the one that interconnects MC and ICT systems at the edge of each system, termed as *edge interface*, and the other that relays information from target MC systems working as data sources to the edge interface and vice versa, termed as *internal interface*. For the former, there are a number of research Brain Machine/Computer Interface (BMI/BCI) [11,12], light-based communication interfaces [4,9,13], implantable [5], which operate as converter between the signals in MC system and the ones in ICT system, while the latter are not well explored.

In this paper, we specifically focus on the internal interface and provide a list of research challenges toward the realization of a generic internal interface. We assume that the internal interface is implemented as other MC systems so that we can deploy them in a self-organized manner, and that we can let them operate autonomously within the human body. In such an internal interface, molecular signals are utilized as common languages as in MC systems. However, molecular signals can be too weak to detect, therefore, signal amplification and relaying are necessary.

The rest of this paper describes an architecture of the IoBNT systems in Sect. 2, and research challenges for internal interface for IoBNT in Sect. 3.

2   An Architecture of the Internet of Bio-Nano-Things (IoBNT) Systems

Figure 1 shows an entire architecture of IoBNT systems. The components of IoBNT systems include MC systems, ICT systems, and interface bridging between MC and ICT systems. MC systems work as a data source. The systems, termed as *target MC systems*, are deployed deep inside the human body to collect physiological data, such as the existence and concentration of specific molecules. It may also be possible to encode physical characteristics in the human body, such as heat, pressure, into molecular signals. MC systems consist of bio-nanomachines communicating information through molecular signals. Each of bio-nanomachines is equipped with an interface that allows the communication between other bio-nanomachines, termed as bio-nanomachine to bio-nanomachine interface (BNI).

ICT systems are responsible for sharing data based on the huge infrastructure of the Internet. The physiological data transmitted from MC systems may be analyzed using machine learning and statistical techniques on the cloud within ICT systems so that the data are interpreted and visualized in an accessible

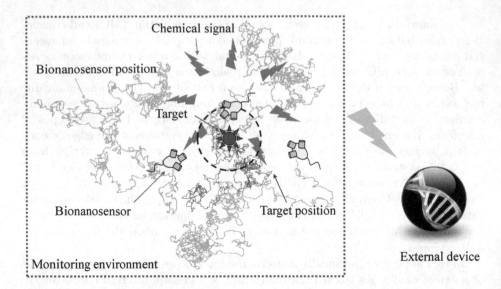

Fig. 1. An architecture of the Internet of Bio-Nano-Things. Adapted from [8].

manner to non-expert users as well as to expert users, such as medical doctors, data scientists.

Figure 2 shows a detailed view of an interface of IoBNT systems. The interface between MC and ICT systems are responsible for converting signals both inside-out and outside-in, which correspond to outmessaging interface (OMI) and inmessaging interface (IMI), respectively [6]. These sub-interfaces, termed as *edge interfaces*, directly communicate with external devices. In this paper, we introduce additional interfaces, termed as *internal interfaces*, that facilitate two-way communication between the edge interfaces and the target MC systems.

3   Research Challenges for Internal Interface for IoBNT Systems

In this paper, we specifically focus on the internal interface between MC and interface MC, and between interface MC and interface MC systems. Molecular signals are utilized as common languages in both directional communication: between other MC systems and interface MC systems, and between external ICT systems and interface MC systems. In both directional communications, molecular signals can be too weak to detect, therefore, (1) signal amplification is inevitable, (2) self-deployment as well as autonomous operation without human intervention is a must, and (3) functionality of relaying information from the target MC and to the external interface is also essential for IoBNT.

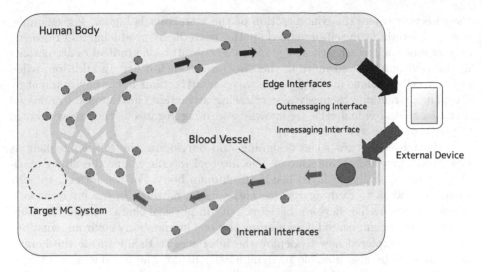

Fig. 2. Interfaces in IoBNT systems.

3.1   Signal Amplification

By utilizing the self-deployment mechanisms discussed in the next subsection, we develop a method for autonomous mobile bionanosensors distributed in space to form self-organized aggregates. By forming macroscopic assemblies of bionanosensors and behaving collectively, we consider implementing functions that are difficult to realize with single bionanosensors. For example, as an interface that interconnects the in vivo and ex vivo environments, a function that converts in vivo molecular signals into electronic signals and a function that amplifies the signal strength will be realized to enable information exchange with existing communication devices. This will enable the interconnection of the bionanosensor network with existing information and communication networks.

Research challenges when designing and developing the signal amplification are what types of signals to use for communication between MC systems. Signals exchanged between MC systems should be determined to neither interfere with nor disturb the operation of other MC systems, especially the target MC system, where molecular signals are exchanged through BNI. On top of that, for the signal amplification, synchronization mechanisms for the aggregated bionanomachines may be also necessary.

3.2   Self-deployment

Forming aggregates is also essential to self-deployment as well as signal amplification. In our work, we proposed a method that mimics the mechanism by which cellular organisms in nature form various spatial structures [7,10]. Many cells in nature use chemotaxis to form various spatial structures. Chemotaxis is the property of a cell to detect the concentration of an attractant and move toward the direction where the concentration of the attractant is higher. For example, certain bacteria form self-organized spatial patterns by producing and releasing attractants, such as aspartic acid, to form a concentration gradient of attractants in the environment and move in the direction of each other. In addition, when starved, cellular slime molds generate waves of attractants in the environment by releasing attractants themselves or releasing attractants in response to released attractants. Individual cells are known to form aggregates as they move toward the source of the wave.

Research challenges when designing and developing the self-deployment is how to deploy internal interfaces. The internal interfaces may be either statically or dynamically deployed inside the human body. For the former, there is theoretical work to form aggregates and specific patterns in the environments where there is neither flow nor heterogeneity in physical characteristics. It could be much more complicated if we assume the real human body environments, but not yet well explored how to deploy the interfaces statically inside the human body. It may be also possible to dynamically deploy the interfaces by making use of blood vessels as the environment. There are several options not limited to the examples explained so far.

3.3   Signal Relay

We proposed a multi-hop drug transport system [14]. The system consists of two types of bionanosensors: a transmitter that emits a signal molecule containing location information to other bionanosensors when it detects a target to which the drug is to be transported, and a receiver that uses the concentration of the signal molecule emitted by the transmitter as a clue to transport the drug to the target. In this system, we proposed to introduce a repeater that implements an infectious propagation scheme to amplify the signal molecules from the transmitter. In the conventional proposed method, the signal molecules are very weak, so the bionanosensors need to be near each other to propagate the signal molecules, and many bionanosensors need to be deployed. By introducing a repeater that implements the infectious propagation method, we showed that the therapeutic agent can be transported efficiently even with a small number of bionanosensors.

By extending this relaying mechanism among bio-nanomachines, we may be able to develop more reliable and functional relaying mechanisms among MC systems. Research challenges in designing and developing the signal relay are what types of signals to use for communication between MC systems, specifically focusing on how to relay signals. What types of signals are well suited for conveying information among MC systems within the human body. When different types of molecules are well-suited for conveying information, the internal interface may be used for converting signaling molecules into the ones.

4   Conclusion

In this position paper, we described research challenges on the Internet of Bio-Nano-Things (IoBNT) specifically focusing on the intermediate interfaces, termed as internal interfaces. The internal interfaces was introduced for the purpose of facilitating two-way communication between the edge interfaces and the target molecular communication systems. We assumed that the target molecular communication systems were deployed deep inside the body, therefore the internal interfaces that make communication between them are essential. This paper described research challenges associated with the internal interfaces.

References

1. Akyildiz, I.F., Pierobon, M., Balasubramaniam, S., Koucheryavy, Y.: The internet of bio-nano things. IEEE Commun. Mag. **53**, 32–40 (2015)
2. Akyildiz, L.F., et al.: PANACEA: an internet of Bio-NanoThings application for early detection and mitigation of infectious diseases. IEEE Access **8**, 140512–140523 (2020)
3. Balasubramaniam, S., Kangasharju, J.: Realizing the internet of Nano things: challenges, solutions, and applications. Computer **46**(2), 62–68 (2013)
4. Ellis-Davies, G.C.R.: Caged compounds: photorelease technology for control of cellular chemistry and physiology. Nat. Methods **4**(8), 619–628 (2007)
5. Kiourti, A., Psathas, K.A., Nikita, K.S.: Implantable and ingestible medical devices with wireless telemetry functionalities: a review of current status and challenges: implantable/ingestible medical devices. Bioelectromagnetics **35**(1), 1–15 (2014)
6. Nakano, T., Kobayashi, S., Suda, T., Okaie, Y., Hiraoka, Y., Haraguchi, T.: Externally controllable molecular communication. IEEE J. Select. Areas Commun. **32**(12), 2417–2431 (2014)
7. Okaie, Y.: Cluster formation by mobile molecular communication systems. IEEE Trans. Mol., Biol. Multi-scale Commun. **5**(2), 153–157 (2019)
8. Okaie, Y., Nakano, T., Hara, T., Nishio, S.: Target Detection and Tracking by Bionanosensor Networks. SpringerBriefs in Computer Science, Springer, Singapore (2016). https://doi.org/10.1007/978-981-10-2468-9
9. Ozawa, T., Yoshimura, H., Kim, S.B.: Advances in fluorescence and bioluminescence imaging. Anal. Chem. **85**(2), 590–609 (2013)
10. Nakano, T., Okaie, Y., Kinugasa, Y., Koujin, T., Suda, T., Hiraoka, Y., Haraguchi, T.: Roles of remote and contact forces in epithelial cell structure formation. Biophys. J. **118**(6), 1466–1478 (2020)
11. Veletic, M., Balasingham, I.: Synaptic communication engineering for future cognitive brain machine interfaces. Proc. IEEE **107**(7), 1425–1441 (2019)
12. Willett, F.R., Avansino, D.T., Hochberg, L.R., Henderson, J.M., Shenoy, K.V.: High-performance brain-to-text communication via handwriting. Nature **593**(7858), 249–254 (2021)
13. Wu, Y.I., Frey, D., Lungu, O.I., Jaehrig, A., Schlichting, I., Kuhlman, B., Hahn, K.M.: A genetically encoded photoactivatable Rac controls the motility of living cells. Nature **461**(7260), 104–108 (2009)
14. Okaie, Y., Ishiyama, S., Hara,T. : Leader-follower-amplifier based mobile molecular communication systems for cooperative drug delivery. In: 2018 IEEE Global Communications Conference (GLOBECOM), pp. 206–212 (2018)

Mathematical Modelling
and Simulations of Biological Systems

Detachment of Microtubules Driven by Kinesin Motors from Track Surfaces Under External Force

May Sweet$^{(\boxtimes)}$ and Takahiro Nitta$^{(\boxtimes)}$ (iD)

Gifu University, Gifu 501-1193, Japan
x3915007@edu.gifu-u.ac.jp, nittat@gifu-u.ac.jp

Abstract. Motor proteins, such as myosin and kinesin, are biological molecular motors involved in force generation and material transport in living cells. Motor proteins and their associated cytoskeletal filaments, such as actin filaments and microtubules, have been utilized for active transport in engineered systems. In controlling the active transport, external forces via electric fields or fluid flow were commonly used. A drawback of using external force is that the external force can cause detachment of microtubules from gliding surfaces. Detachment leads to loss of cargo or sparse surface density of microtubules, thus limiting the availability of external forces. Detachment should be minimized. In doing so, detailed observation on the process of detachment would be helpful. However, due to its limited spatial and temporal resolution, experimental investigations are hampered. Here, we show a simulation study for the detachment of microtubules gliding over surfaces coated with kinesin motors by an external force. Owing to the computer simulation's high spatial and time resolution, two modes of detachment were found. Detailed processes of the two modes were revealed, which would be useful to diminish detachment.

Keywords: Biomolecular motor · Cytoskeletal filaments · Computer simulation

1 Introduction

Motor proteins, such as myosin and kinesin, are biological molecular motors working for force generation and intracellular transport. In the last two decades, they have been utilized in engineered systems [1]. Molecular shuttles driven by motor proteins are essential for active transport in such devices, where cytoskeletal filaments carrying analytes [2–4] or information [5, 6] are transported along predefined tracks covered with associated motor proteins to designated destinations.

To control the movements of molecular shuttles, external forces via electric fields or fluid flow were commonly used [7–9]. While the larger applied force leads to better control, the drawback is the detachment of molecular shuttles from surfaces over which they glide [8]. The detachments lead to loss of cargo and the sparse surface density of molecular shuttles [10], which limits device performance and should be minimized.

T. Nakano (Ed.): BICT 2021, LNICST 403, pp. 199–206, 2021.
https://doi.org/10.1007/978-3-030-92163-7_16

In addition, while trajectories without detachment can be predicted with computer simulations [11, 12], detachment makes such prediction difficult, bringing complications in designing devices. A detailed mechanism of the detachment of molecular shuttles would be helpful to their suppression. However, the process of detachment is abrupt compared to spatial and temporal limitations in experiments so that detailed mechanisms of detachment remain unknown.

Here, to investigate detailed mechanisms of the detachment of the molecular shuttles, we used our own developed computer simulation. The use of the computer simulation enables us to reveal a detailed mechanism of the detachment, which cannot be obtained in experimental investigation. In the present study, we found that there are two distinct modes of microtubule detachment: unzipping/swiveling and jumping. Owing to the high time resolution of the simulation, we observed the two detachment modes with around 1,000 times higher time resolution than that of a conventional microscope. Our observation will guide the detailed understanding of the mechanism of the detachment of molecular shuttles driven by the kinesin motor.

2 Simulation Method

The simulation method was developed based on a previous study [13] and is briefly described as follows.

Microtubules were modeled as inextensible elastic beams with a bending rigidity of $22 \, \text{pN}/\mu\text{m}^2$ subjected to thermal fluctuations and applied force, represented by bead-rod polymers. The applied force acting on microtubules was represented by the force acting on the beads. The length of the microtubules was $5 \, \mu\text{m}$, compatible with the length of microtubules commonly used in experiments.

Microtubules were propelled by kinesin motors, which are modeled as a linear spring with a spring constant of $100 \, \text{pN}/\mu\text{m}$ and with zero equilibrium length. The heads of the kinesin motors moved toward the microtubule plus end, while the tails of kinesin motors were fixed on the substrate, building up tension to move the microtubules toward their minus ends.

The equations of motion of beads consisting of microtubules were expressed to be

$$r'_i(t + \Delta t) = r_i(t) + \frac{\Delta t}{\zeta}\left(F_{b,i} + F_{k,i} + F_{ex,i}\right) + \sqrt{2D \cdot \Delta t} \cdot \xi_i,$$

where ζ is the viscous drag coefficient, $F_{b,i}$ is the force due to bending of microtubules, $F_{k,i}$ is the force exerted by kinesin motors, $F_{ex,i}$ is an external force, ξ_i is a three-dimensional random vector whose components take random values with zero mean and standard deviation of one. The last term of the right-hand side of the equation represents the thermal fluctuation of the beads. The calculated positions, $\{r'_i(t + \Delta t)\}$, were corrected to have an assigned distance of $0.25 \, \mu\text{m}$ between them.

3 Results

3.1 Trajectories Under High External Forces

Figure 1 shows representative trajectories of microtubules gliding over kinesin motors with a surface motor density of $10 \, \mu\text{m}^{-2}$ under an external force of $5.0 \, \text{pN}/\mu\text{m}$.

The trajectories showed intermittencies reflecting partial or whole microtubule detachment from the surface. In contrast, trajectories with higher motor density and/or weaker external force did not show such intermittencies. Since the partial and whole microtubule detachments occurred stochastically, the control of microtubule movement was diminished.

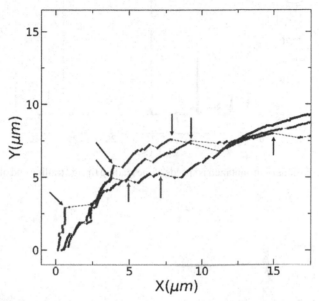

Fig. 1. Representative trajectories of microtubules gliding over kinesin motors with a surface motor density of $10 \, \mu m^{-2}$ under an external force of 5.0 pN/μm. Different colors represent different microtubule trajectories. Jumps are denoted with arrows. Only major jumps are marked for visibility.

To quantify the intermittencies in the trajectories, the instantaneous speed of leading tips and trailing ends were calculated (Fig. 2). Reflecting the partial and whole microtubule detachments, the time evolutions of the instantaneous speeds showed sharp isolated peaks over the average speed of 0.8 μm/s. The instantaneous speeds at the peaks could reach more than 50 times the average speed of 0.8 μm/s. The instantaneous speed showed an exponential distribution. No significant difference was observed between distributions of the leading tips and trailing ends (Fig. 3). This observation indicated that detachments equally occurred from both the leading tips and trailing ends.

We found that detachment processes can be categorized into two classes: unzipping/swiveling (Fig. 4a) and jumping (Fig. 4b). We will describe each process and mechanism in the following sections.

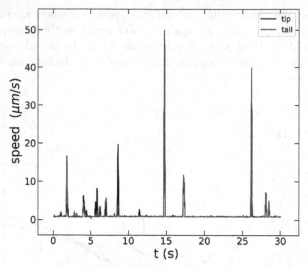

Fig. 2. Time evolutions of instantaneous speeds of the leading tip and trailing end of a microtubule.

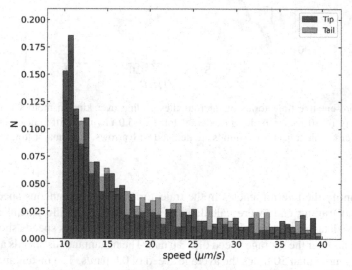

Fig. 3. Histograms of instantaneous speeds of the leading tips and trailing ends of 60 microtubules. To highlight the abrupt changes in gliding speed corresponding to the peaks in Fig. 2, only the part with the speed >1.0 μm/s is shown.

3.2 Unzipping/Swiveling of Microtubule

Unbinding of kinesin motors from microtubules successively occurred from either the leading tip or the trailing end, which we call unzipping. A representative microtubule undergoing unzipping followed by swiveling is shown in Fig. 4a observed with 0.1 s time resolution. The 0.1 s time resolution could not clearly identify the phenomenon

Fig. 4. Unzipping/swiveling (a) and jump (b) observed with 0.1 s time resolution. The orange lines represent superimposed conformations of a microtubule. White dots represent kinesin motors. Green dots represent kinesin motors binding to the microtubule. (Color figure online)

of microtubules detachment. To identify the detailed phenomenon of microtubules, we magnified with 1,000 times higher time resolution in Fig. 5. Part of the microtubules showed lateral shifts a few times. Upon unzipping, the population of binding kinesin motors was gradually replaced. Between the shifts, microtubules showed thermal fluctuations without changing the population of binding kinesin motors. After unzipping, swiveling of microtubules occasionally followed, which caused further changes in their gliding direction.

While unzipping was equally likely to occur from both ends, unzipping from leading tips affected trajectories more significantly than that from the trailing ends, since trajectories are mostly determined by their leading tips. Unzipping/swiveling tended to occur more often when the microtubules were moving perpendicular to an applied force.

3.3 Jump of Microtubule

In contrast to unzipping/swiveling, the jumps abruptly occurred within 0.1 s time resolution. In many cases, the positions of microtubules just shifted downstream of the applied force without significant change of direction. A representative microtubule undergoing jump is shown in Fig. 6 observed with 0.1 ms time resolution. Upon jump, the whole population of binding kinesin motors was replaced. Distance during the jump varies and can be more than the length of microtubules. The jumps observed with a 0.1 s time window consisted of smaller jumps without changing the direction of microtubules.

The jumps of microtubules were observed with almost equal frequency all the way along the trajectories. The jump of microtubules was greatly diminished by lowering the applied force from 5.0 pN/μm to 4.0 pN/μm.

Fig. 5. Unzipping and swiveling observed with 0.1 ms time resolution. The orange lines represent superimposed conformations of a microtubule. White dots represent kinesin motors. Green dots represent kinesin motors binding to the microtubule. (Color figure online)

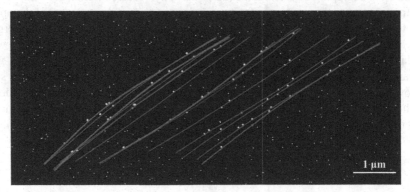

Fig. 6. A series of small jumps observed with 0.1 ms time resolution. The orange lines represent the superimposed conformations of a microtubule. White dots represent kinesin motors. Green dots represent kinesin motors binding to the microtubule. (Color figure online)

4 Discussion

By using our own developed computer simulation, we investigated the detailed mechanism of the detachment of molecular shuttles driven by kinesin motors. We found that there were two modes of detachment, unzipping/swiveling and jumping. Unzipping/swiveling can be characterized by the sequential unbinding of kinesin motors from microtubules from either the leading tips or the trailing ends. On the other hand, jumping can be characterized by sudden shifts of microtubule position toward the downstream of the external force without significant change of their direction. These kinds of observations were not obtained before because of limited space and time resolution in experiments.

Based on our observations reported here, to diminish the detachment of microtubules from track surfaces, one should take note of the following two. Firstly, the applied force should be lower than the force inducing a significant occurrence of jump of microtubules. Secondly, to avoid unzipping/swiveling of microtubules, a strong external force should not be applied when the microtubules are moving perpendicularly to the force.

References

1. Saper, G., Hess, H.: Synthetic systems powered by biological molecular motors. Chem. Rev. 120(1), 288–309 (2020). https://doi.org/10.1021/acs.chemrev.9b00249
2. Lin, C.T., Kao, M.T., Kurabayashi, K., Meyhofer, E.: Self-contained, biomolecular motor-driven protein sorting and concentrating in an ultrasensitive microfluidic chip. Nano. Lett. 8(4), 1041–1046 (2008). https://doi.org/10.1021/nl072742x
3. Fischer, T., Agarwal, A., Hess, H.: A smart dust biosensor powered by kinesin motors. Nat. Nanotechnol. 4(3), 162–166 (2009). https://doi.org/10.1038/nnano.2008.393
4. Lard, M., et al.: Ultrafast molecular motor driven nanoseparation and biosensing. Biosens. Bioelectron. 48, 145–152 (2013). https://doi.org/10.1016/j.bios.2013.03.071
5. Nakano, T., Moore, M.J., Wei, F., Vasilakos, A.V., Shuai, J.: Molecular communication and networking: opportunities and challenges. IEEE Trans. Nanobioscience 11(2), 135–148 (2012). https://doi.org/10.1109/TNB.2012.2191570
6. Farsad, N., Yilmaz, H.B., Eckford, A., Chae, C.B., Guo, W.: A comprehensive survey of recent advancements in molecular communication. IEEE Commun. Surv. Tutorials 18(3), 1887–1919 (2016). https://doi.org/10.1109/COMST.2016.2527741
7. Van Den Heuvel, M.G.L., De Graaff, M.P., Dekker, C.: Molecular sorting by electrical steering of microtubules in kinesin-coated channels. Science 312(5775), 910–914 (2006). https://doi.org/10.1126/science.1124258
8. Agayan, R.R., et al.: Optimization of isopolar microtubule arrays. Langmuir 29(7), 2265–2272 (2013). https://doi.org/10.1021/la303792v
9. Isozaki, N., Shintaku, H.,Kotera, H., Hawkins, T.L., Ross, J. L., Yokokawa, R.: M I C R O R O B O T S Control of molecular shuttles by designing electrical and mechanical properties of microtubules. [Online]. Available http://robotics.sciencemag.org/ (2017)
10. Bassir Kazeruni, N.M., Rodriguez, J.B., Saper, G., Hess, H.: Microtubule detachment in gliding motility assays limits the performance of kinesin-driven molecular shuttless. Langmuir. 36(27), 7901–7907 (2020). https://doi.org/10.1021/acs.langmuir.0c01002
11. Nitta, T., Tanahashi, A., Hirano, M., Hess, H.: Simulating molecular shuttle movements: Towards computer-aided design of nanoscale transport systems. Lab. Chip. 6(7), 881 (2006). https://doi.org/10.1039/b601754a

12. Nitta, T., Tanahashi, A., Hirano, M.: In silico design and testing of guiding tracks for molecular shuttles powered by kinesin motors. Lab. Chip. **10**(11), 1447 (2010). https://doi.org/10.1039/b926210e
13. Ishigure, Y., Nitta, T.: Understanding the Guiding of Kinesin/Microtubule-Based Microtransporters in Microfabricated Tracks. Langmuir **30**(40), 12089–12096 (2014). https://doi.org/10.1021/la5021884

A Mathematical Model Predicting Gliding Speed of Actin Molecular Shuttles Over Myosin Motors in the Presence of Defective Motors

Samuel Macharia Kang'iri[✉] and Takahiro Nitta[✉]

Gifu University, Gifu 501-1193, Japan
x3914101@edu.gifu-u.ac.jp, nittat@gifu-u.ac.jp

Abstract. Motor proteins are molecular machines that operate in living cells. These motor proteins have been used in vitro for applications such as nano- and microscale devices as transport systems in biosensors, biocomputing, and molecular communication. By introducing motor proteins into these devices, motor proteins become defective due to unfavorable binding to device surfaces, causing a decrease in transport speed or malfunctioning of transport. However, systematic experimental investigations of the effects of defective motors are hampered by difficulties in controlling the number of defective motors on surfaces. Here, we show a systematic study on the effects of defective motors on the motility of transport by using a mathematical model. The model predicted that motility is independent of the length of the associated filaments and depends on the ratio of the active motors. The model revealed that the ratio of active motors of more than 80% is required for sustainable motility. This insight would be useful in choosing appropriate materials for devices integrated with motor proteins.

Keywords: Biomolecular motor · Cytoskeleton · Biosensor

1 Introduction

Motor proteins are molecular machines that operate in living cells. These motor proteins include myosin and kinesin, which travel along the actin filament and microtubule, respectively. Through a cycle of filament binding and release, these motor proteins convert chemical energy directly to mechanical work for material transport and actuation in living cells. These motor proteins have been used in vitro for applications such as nano- and microscale devices [1]. Molecular shuttles (MS) are based on motor proteins and are essential for active transport. They have been implemented in biosensors [2–4], biocomputation [5], and molecular communication [6, 7].

As the implementation proceeds, practical problems such as the effects of defective motors arise. Defective motors, sometimes called "dead" head, ATPase catalytic domain, bind to cytoskeletal filaments but are unable to hydrolyze adenosine triphosphate (ATP), thus serving as an effective impedance to the translocation of cytoskeletal filaments driven by active motors. This causes fishtailing, swirling, and halting of the filaments [8].

T. Nakano (Ed.): BICT 2021, LNICST 403, pp. 207–214, 2021.
https://doi.org/10.1007/978-3-030-92163-7_17

While actin filament (AF) and myosin-based MS is preferable in terms of gliding speed than microtubule and kinesin-based one [9], it is more susceptible to the "dead" heads. To achieve smooth movements of AFs gliding over myosin motors in in vitro motility assay, "dead" motors have to be carefully removed prior to observation of movement, and the surfaces should be passivated to prevent "non-ideal" adhesion of myosin motors to the surface leading to denature [10]. Such coatings may not be available for use in applications since in biophysical studies, photoresists are commonly used instead of glass. Hansson *et al.* investigated the effects of various polymer materials commonly used for microfabrication on actin motility, and some fractions of myosin motors became defective depending on the nature of the polymer materials, such as hydrophobicity [11]. However, a limitation of such experimental approaches arises from the difficulty in controlling the precise amount of defective motors on surfaces, which hampers the systematic investigation.

Here, to provide predictions of gliding speed on AFs driven by myosin motors in the presence of defective motors, we developed a mathematical model. In contrast to the experimental work, the mathematical model enables a systematic quantitative investigation of the effects of defective motors on the motility of AFs. The simplicity of the mathematical model makes it easy to gain insights into the effect of defective motors.

2 Mathematical Modelling

To understand the underlying mechanism of the translocation impedance, we developed a 1D analytical model based on a previous study [12]. Our 1D model assumes that an AF is propelled by active myosin against impedance by defective ones with a gliding speed (v) (Fig. 1). The gliding speed was assumed to depend on the average force acting on each active myosin (f):

$$v = v_{max}\left(1 - \frac{f}{f_{stall}}\right) \tag{1}$$

where v_{max} is the maximum speed of the actin translocation, and f_{stall} is the stall force (-0.4 pN) [13, 14].

The acting force, f, is assumed to be exerted by the defective myosin. During the AF translocation, defective motors undergo repeated cycles of association with an AF, elongation and dissociation (Fig. 2). When an AF comes close, a defective myosin binds to the AF with a rate of $1/\tau_1$. Once bound, the defective motor is stretched by the AF translocation, building up tension impeding the AF translocation. When the tension reaches the rupture force of myosin, f_{rupt} (-9.2 pN) [15], the defective myosin dissociates from the AF. For simplicity, spontaneous dissociation of defective myosin from AF was neglected. The duration that the defective myosin binds to the AF, τ_2, depends on the AF speed and is given by $\tau_2 = -f_{rupt}/kV$, where k is the spring constant of the defective motor. Thus, the time-averaged friction force generated by a defective myosin, $\overline{f_{def}}$, is given by

$$\overline{f_{def}} = \frac{1}{\tau_1 + \tau_2} \int_0^{\tau_1 + \tau_2} f_{def}\, dt. \tag{2}$$

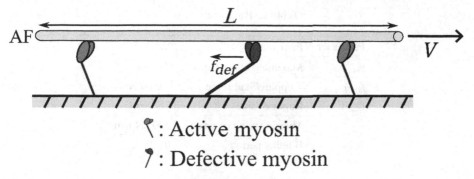

$\m{:}$: Active myosin

$\mathit{?}$: Defective myosin

Fig. 1. A schematic drawing of actin molecular shuttles over myosin motors in the presence of defective motors.

Guided by Fig. 2, the integral can be calculated, leading to

$$\overline{f_{def}} = \frac{f_{rupt}}{2(1 - kv\tau_1/f_{rupt})}. \tag{3}$$

Since the number of the active myosins binding to the AF with the length of L is $\rho_a L$ and that of the defective ones $\rho_d L$, where ρ_a and ρ_d are the line densities of active and defective myosins, respectively, the impedance per active myosin is given by

$$f_{imp} = \frac{\rho_d}{\rho_a} \frac{f_{rupt}}{2(1 - kv\tau_1/f_{rupt})}. \tag{4}$$

The parameters used in this study are shown in Table 1. To obtain numerical solution, we used MATLAB.

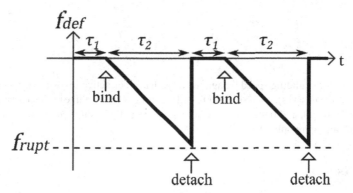

Fig. 2. A schematic representation of the time evolution of the force generated by a defective myosin.

Table 1. Parameters used

Parameter	Particulars	Value
v_{max}	Maximum gliding speed	7 μm/s
f_{stall}	Stopping/Stall force	−0.4 pN
f_{rupt}	Rupture force	−9.2 pN
k	Myosin spring constant	300 pN/μm
τ_1	Binding period	0.025

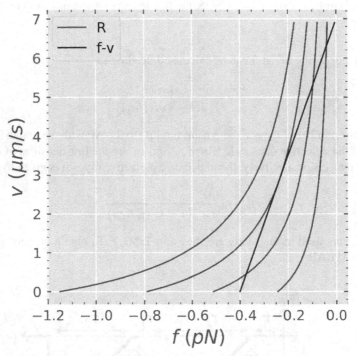

Fig. 3. A plot of the gliding speed, v, against impedance per active myosin with various active motor ratios, R (red curves), overlaid with f-v relation: Eq. (1) (blue line). The active motor ratio is 0.800, 0.854, 0.900, and 0.950 from left to right. The critical active motor ratio, in this case, was 0.854 (Color figure online).

3 Results

From Eq. (1), once the impedance is given, the AF gliding speed can be calculated. On the other hand, to determine the impedance from Eq. (3), the AF gliding is needed. Thus, to obtain the AF gliding speed, we need to solve Eqs. (1) and (3) in a self-consistent manner. That is, the gliding speed is given by the intersection of the Eqs. (1) and (3)

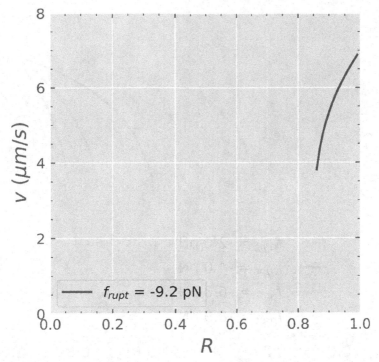

Fig. 4. The AF gliding speed, v, as a function of the active motor ratio, R.

(Fig. 3). The single parameter to determine the intersection is the gliding speed, ρ_d/ρ_a, and is independent of the length of AF.

Depending on ρ_d/ρ_a, there are three cases. At small ρ_d/ρ_a (the red curve for the active motor ratio of 0.950 in Fig. 3), there is only one solution. At an intermediate (the red curve for the active motor ratio of 0.900 in Fig. 3), there are two solutions. Above the critical value (the red curve for the active motor ratio of 0.800 in Fig. 3), there is no solution. In the case that there are two solutions, the one with the higher gliding speed is stable while the other is unstable as described below. Firstly, we consider fluctuations of the impedance around the solution with a higher gliding speed. Although we have so far discussed an averaged behavior, force fluctuation occurs due to continuous and stochastic association and dissociation of defective motors. For example, when AF happens to be decelerated by fluctuation, the impedance at this decreased speed is smaller, resulting in accelerating the AF back to the original gliding speed, showing that the solution is stable. On the other hand, around the solution with the lower gliding speed, when AF happens to be decelerated by fluctuation, the impedance at this decreased speed is larger, leading to the further deceleration of the AF, showing that the solution is unstable.

Taking the stable solutions, the gliding speed was obtained as a function of the active motor ratio, defined as $\rho_a/(\rho_a + \rho_d)$ (Fig. 4). The gliding of AFs can only occur with a rather high active motor ratio of 0.854 or more. Interestingly, due to the instability, the

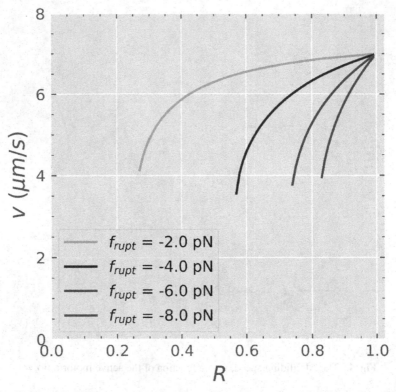

Fig. 5. The AF gliding speed as a function of the active motor ratio with various f_{rupt}.

gliding speed did not continuously increase at the onset of gliding but showed an abrupt change from 0 to 4 μm/s.

Since we neglected the spontaneous dissociation of defective myosin from AF, our choice of the rupture force of −9.2 pN may be an overestimate. We investigated the AF gliding speed as a function of the active motor ratio with various rupture forces ranging from −2.0 pN to −8.0 pN. The lower the rupture force, the lower the threshold active motor ratio for the onset of gliding movements. The abrupt change in gliding speed at the onset of motility persisted (Fig. 5).

4 Discussion

Our analytical model showed that the active motor ratio $\rho_a/(\rho_a + \rho_d)$ is the single important factor that influences the actin filament speed and that an active ratio of more than 80% is required for continuous gliding movement. The need for a high critical active motor ratio for gliding is consistent with the fact that procedures to remove defective heads are needed to achieve consistent gliding for actin/myosin in vitro motility assay.

This finding suggests that the substrate material should be carefully selected such that denaturing of the adhering myosin motors is minimized. In a study conducted by Hanson et al. [11], higher material hydrophobicity was found to be associated with increased HMM surface adsorption but lowers the number of active HMM molecules. This study suggested that the preferable materials are those that retain the activity of motors even though the total amount of motors are limited.

The force-velocity given by Eq. (1) was employed for simplicity, which was helpful to obtain an analytical solution of the critical active motor ratio. Using a more realistic relationship such as the Hill equation [16] may yield a more accurate prediction with a drawback due to the complexity of the mathematical expression. Another shortcoming of this study would be that only 1D movements were considered. The use of computer simulation [17] would be a complementary approach to this study by dealing with 3D movements. In addition, systematic experiments with various mixing ratios of myosin and mutated defective myosin would enable qualitative validation of our mathematical model, although the exact active motor ratio would be difficult to obtain due to uncontrollable conversion of active to defective myosin.

References

1. Saper, G., Hess, H.: Synthetic systems powered by biological molecular motors. Chem. Rev. **120**(1), 288–309 (2020)
2. Lin, C.T., Kao, M.T., Kurabayashi, K., Meyhofer, E.: Self-contained, biomolecular motor-driven protein sorting and concentrating in an ultrasensitive microfluidic chip. Nano Lett. **8**(4), 1041–1046 (2008)
3. Fischer, T., Agarwal, A., Hess, H.: A smart dust biosensor powered by kinesin motors. Nat. Nanotechnol. **4**(3), 162–166 (2009)
4. Lard, M., et al.: Ultrafast molecular motor driven nanoseparation and biosensing. Biosens. Bioelectron. **48**, 145–152 (2013)
5. Nicolau, D.V., et al.: Parallel computation with molecular-motor-propelled agents in nanofabricated networks. Proc. Natl. Acad. Sci. USA **113**(10), 2591–2596 (2016)
6. Farsad, N., Yilmaz, H.B., Eckford, A., Chae, C.-B., Guo, W.: A comprehensive survey of recent advancements in molecular communication. IEEE Commun. Surv. Tutorials **18**(3), 1887–1919 (2014)
7. Nakano, T., Moore, M.J., Wei, F., Vasilakos, A.V., Shuai, J.: Molecular communication and networking: opportunities and challenges. IEEE Trans. Nanobiosci. **11**(2), 135–148 (2012)
8. Bourdieu, L., Duke, T., Elowitz, M.B., Winkelmann, D.A., Leibler, S., Libchaber, A.: Spiral defects in motility assays: a measure of motor protein force. Phys. Rev. Lett. **75**(1), 176–179 (1995)
9. Nitta, T., et al.: Comparing guiding track requirements for myosin- and kinesin-powered molecular shuttles. Nano Lett. **8**(8), 2305–2309 (2008)
10. Rahman, M.A., Salhotra, A., Månsson, A.: Comparative analysis of widely used methods to remove nonfunctional myosin heads for the in vitro motility assay. J. Muscle Res. Cell Motil. **39**(5–6), 175–187 (2019)
11. Hanson, K.L., et al.: Polymer surface properties control the function of heavy meromyosin in dynamic nanodevices. Biosens. Bioelectron. **93**, 305–314 (2017)
12. Greenberg, M.J., Moore, J.R.: The molecular basis of frictional loads in the in vitro motility assay with applications to the study of the loaded mechanochemistry of molecular motors. Cytoskeleton **67**(5), 273–285 (2010)
13. Kishino, A., Yanagida, T.: Force measurements by micromanipulation of a single actin filament by glass needles. Nature **334**(6177), 74–76 (1988)

14. Riveline, D., et al.: Acting on actin: The electric motility assay. Eur. Biophys. J. **27**(4), 403–408 (1998)
15. Nishizaka, T., Miyata, H., Yoshikawa, H., Ishiwata, S., Kinosita, K.: Unbinding force of a single motor molecule of muscle measured using optical tweezers. Nature **377**(6546), 251–254 (1995)
16. Hill, A.V.: The heat of shortening and the dynamic constants of muscle. Proc. R. Soc. London Ser. B Biol. Sci. **126**(843), 136–195 (1938)
17. Ishigure, Y., Nitta, T.: Simulating an actomyosin in vitro motility assay: toward the rational design of actomyosin-based microtransporters. IEEE Trans. Nanobiosci. **14**(6), 641–648 (2015)

Limits of Intelligence and Design Implication

Son Tran[✉] , Sophie Alyx Taylor , and Dan V. Nicolau Jr.

School of Mathematical Sciences, Queensland University of Technology, Brisbane, Australia
caoson.tran@qut.edu.au

Abstract. This paper presents a design framework for artificial general intelligence (AGI). The approach is guided by a simple question: if we encounter an intelligent system, what could we observe? The answer is based on the idea that intelligence emerges from simple goal-driven interactive adaptability, and the process leads to emerging properties underlying complex behaviors of any intelligent systems. These properties in turn serves as design criteria for the construction of a network architecture of AGI which is proposed here for further investigations in future studies.

Keywords: Artificial general intelligence · Cognitive redundancy · Cognitive complexity · Administrative behaviors · Composition search · Network of mind · Hypergraph categories

1 Emerging Properties of Intelligence

What could we learn about intelligence? Simon [1] approached this question by introducing the concept of interface to study intelligent systems. It is essentially a collection of properties emerging out of the system's behaviors seeking to achieve specific goals. These properties in turn impose limits on what the system could or could not do. Using the interface, an observer could predict how the system behaves when facing specific events without knowing details of its internal reasoning processes.

Minsky [2] proposed another goal-based framework focusing instead on constructing internal processes underlying an intelligent system's behaviors. In this framework, the system's intelligence relies on a set of resources that could be combined in various ways to create processes corresponding to different behaviors. A central concept is resource management which deals with how resources, diverse in their representation and capability, are acquired, evaluated, allocated, combined, and used. The system's intelligence could then be observed through these administrative operations.

In this section, these concepts are integrated into a simple computational model to explore the idea that intelligence is closely connected to computational complexity created by the process of continuous managing resources to achieve goals. By looking at the interface of this process, the model discusses three emerging properties of intelligence: cognitive redundancy, cognitive complexity, and administrative behaviors.

T. Nakano (Ed.): BICT 2021, LNICST 403, pp. 215–229, 2021.
https://doi.org/10.1007/978-3-030-92163-7_18

1.1 Cognitive Redundancy and Complexity

The computational model views an intelligent system's behaviors as goal-driven processes that use resources to achieve goals when encountering specific events. These processes are represented by the following structure.

$$\{e \rightarrow g \rightarrow r\} \tag{1}$$

where e is an *event situation* which describes an event, g is a *goal situation* that describes the goals the system wants to achieve given e, and r is a *resource situation* describing resources the system uses to achieve goals. $\rightarrow$ is an operator that maps one type of operator to another type of operator. The detailed construct of e, g, r, and $\rightarrow$ can be ignored here since the abstraction is sufficient for the purpose of this paper. In other words, we only look at the situations' interfaces. Essentially (1.1.1) says that when an event happens, the system decides whether it should set goals or not. Given a set of goals, it proceeds to figure out the resources needed and how they could be used together to achieve the goals.

What could the tiny structure in (1.1.1) tells us about intelligence? The most obvious thing is that it describes an ideal situation in which we always know precisely what goals to set and what resources to use. But the picture is less clear when we have a circumstance in which event, goal, and resource situations interact to create complexity. Consider the following three scenarios.

$$\{\{e \rightarrow e_1\} \rightarrow^e \{\{e \rightarrow e_1\}, \{e \rightarrow e_2\}, \{e_1 \rightarrow e_2\}\}\} \tag{2}$$

$$\{\{g \rightarrow g_1\} \rightarrow^g \{\{g \rightarrow g_1\}, \{g \rightarrow g_2\}, \{g \rightarrow g_3\}\}\} \tag{3}$$

$$\{\{\{r \rightarrow r_1\}, \{r \rightarrow r_2\}\} \rightarrow^r \{\{r_1 \rightarrow r_2\}, \{r \rightarrow r_3\}, \{r_1 \rightarrow r_3\}, \{r_2 \rightarrow r_3\}\}\} \tag{4}$$

The structure in (2) could be described in the following process. First, the event situation e first encountered by the system leads to another event situation e_1. This in turn leads to a new *event space* created in the system's memory to store *configurations* in the form $\{e \rightarrow e_1\}$. The system now learns that there are more possible configurations of the event space, a realization represented by the structure on the right of (2). The updating process is represented by the operator $\rightarrow^e$ constructed as a term-rewriting operation on hyper graphs [3–5]. Similar scenarios could be observed for the *goal space* in (3) and the *resource space* in (4). The structure in (1) now expands into:

$$\{\{e, \rightarrow^e\} \rightarrow \{g, \rightarrow^g\} \rightarrow \{r, \rightarrow^r\}\}, \tag{5}$$

with $\{e, \rightarrow^e\}$, $\{g, \rightarrow^g\}$, and $\{r, \rightarrow^r\}$ representing (2), (3), and (4) correspondingly.

The system's ability to update event, goal, and resource spaces creates a cognitive redundancy that imposes a limit on how flexible the system could be in term of creating connections and switching between configurations to adapt to changes. Figure 1 (left panel) visualizes the event space as a hyper graph, showing how diverse it has become just after 6 updates from the initial configuration in (2). This leads to an increase in cognitive redundancy observed in the rapid raise in the number of edge counts (Fig. 1, right panel). From this perspective, the concept of cognitive redundancy could explain why models based on artificial neural networks, while performing well in some specific tasks, often failed or perform poorly in tasks found simple by humans [6]. The reason is that these models have rather low cognitive redundancy since their architectures are essentially extensions of (1) which has only one updating path.

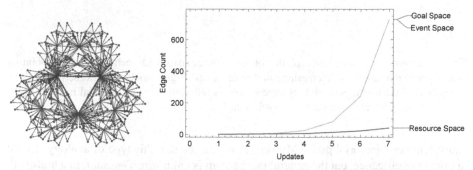

Fig. 1. Left panel: the hypergraph of the event space after 6 updates from the simple initial configuration $\{e \rightarrow e_1\}$. Right panel: an increase in cognitive redundancy as showed by a rapid increase in the number of edge count.

Cognitive redundancy, however, comes with a cost. With a space of many possible updating paths, the system must essentially perform a combinatorial task to select the desired path to act upon. Besides, the combinatorial space is much larger than that created by the hypergraph's edges since the graph does not reflect possible updating paths, instead it simply merges them into a reduced form. As a result, searching for a right combination of resources situations requires exploration of all possible path absent of effective administrative operations (Fig. 2, left panel). This combinatorial challenge leads to *cognitive complexity*, which corresponds to an exponential increase in the number of possible updating paths created by simple updating rules (Fig. 2, right panel). Cognitive redundancy and cognitive complexity are thus emerging properties that impose constraints on the system's ability to find paths to its goals.

1.2 Administrative Behaviors

What should the system do when facing cognitive complexity? A likely response is to seek for a balance between redundancy and complexity to achieve desired goals. This implies that there should be a mechanism to achieve this balance, which is essentially an administrative task. The most obvious thing for the system to do is to impose a set of constraints on the scope of the updating process. For example, it could impose limits on the number of updating steps or the type and scope of resource situation. Alternatively, it could simply restrict its attention to specific event, goal, and resource situations. For example, an organization could create an identity, focus on specific goals and work within specified budget plans to manage complexity [7].

The multipath evolution of the system hints at a more elaborate mechanism involving two phases. First, it could adopt a kind of analogical reasoning that could rapidly identify

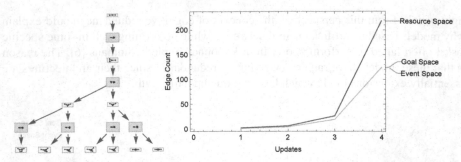

Fig. 2. Left panel: the multipath evolution of the resource space with each possible path pointing to a hyper graph having an increasing number of edges. Right panel: an increase in cognitive complexity in the event, goal, and resource spaces as reflected in the exponential increase in the number of edge counts in the multipath evolution.

equivalent combinatorial paths. Minsky [2] suggested that this type of analogy is used in human intelligence, but the general mechanism is unknown. That said, in a multipath system, analogical reasoning could be viewed as performing an operation similar to the Knuth-Bendix completion algorithm that seeks to reduce two updating paths to the same structure (Fig. 3) [5, 8]. Another possibility is the deployment of special type of resources that are essentially computational systems specializing in doing combinatorial tasks quickly. These resources thus help reduce the search space to a more manageable structure. An example is the use of a hybrid physical-biological computation network to perform massive parallel computing tasks [9].

In the second phase, the system performs domain specific credit assignment operations which are essentially weight assignment tasks. An intelligent system is thus likely to have a memory of different credit assignment methods corresponding to different configuration of event, goal, and resource situations [2]. And this implies acquisition of a large memory of decision processes made by diverse intelligent systems.

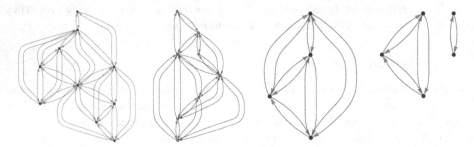

Fig. 3. Left panel: Knuth-Bendix completion rules for the event space. Middle panel: Knuth-Bendix completion rules for the goal space. Right panel: Knuth-Bendix completion rules for the resource space.

The model proposed here is far from a complete and accurate description of intelligence at an abstract level. It presents a computational way to investigate properties of intelligence that has novel architectural implications. The hypergraph structure with dynamic updating rules enables us to model intelligence at a level sufficiently abstract to accommodate different problem domains. Through the model, we discovered two important concepts, cognitive redundancy and cognitive complexity, and this in turn leads to the emergence of administrative behaviors as key mechanisms underlying the evolution of intelligence. Using these constructs as design criteria, the next section presents a design framework called network of mind (NoM) to demonstrate how complex intelligent systems could emerge from the simple task of dealing with redundancy and complexity during the search for paths to their goals.

2 The Architecture of NoM

Several conceptual constructs are used in NoM to represent three ideas. First, intelligence is simply considered as a process in which a system learns how to achieve goals by trying to be resourceful given specific circumstances. Being resourceful here means having the ability to combine and switch among different ways of doing things. Second, the learning process arises not only from internal capability but also external interaction with other systems. Finally, intelligence processes are implemented in a network structure that makes difference between layers of representation of knowledge rather superfluous while enabling massive parallel operations to deal with cognitive complexity in a fashion inspired by Nicolau et al. [9].

2.1 Key Constructs

Imagine a world in which things exist in construct called *entities*. Each entity is associated with a finite collection of *interfaces* through which it interacts with other entities. The interaction process takes place when an interface receives some input and sends out output. The interaction is feasible when the output sent out by one entity are accepted as input for at least another entity. Input and output could themselves be entities. Entities that do not have an interface are called uncertainties.

$$\text{Entity} = \{\{interface \rightarrow \{input, output\}\}\}_1^n \tag{6}$$

Some entities send out their output through the interface without having definite receiving destinations, while others deliberately send output to specific targets with the expectation of getting something back. An entity behaving like the latter are called an agent. Its interaction with other entities is called a probing process in which the agent uses a set of entities called models to extract information from the other entities through their interfaces. A model's output could be used as another entity's input, and the entity's output another model's input. A model, however, only works through a single interface, and an agent could use different models to probe the same interface of an entity. The agent also needs to update its models when no longer relevant or less effective.

$$\text{Model} = \{interface \rightarrow \{input, output\}\} \tag{7}$$

Each agent only pays attention to certain entities called entities of interest (EOI). The probing models thus play a dual role. First, they serve as a tool for the agent to discover an EOI's interfaces. Second, they are used as learning mechanisms that enable the agent to extract properties associated with the EOIs in a probing process. A property is essentially the result of a transformation of a probing process' output into constructs that remain invariant to the agent. From time to time an interface's output could be used as properties without further transformation. Different models learn about different properties of the same EOI.

When the agent wants to make changes to its EOI's properties, it needs to use some models to interact with the EOI's interfaces. This is called an affecting process in which the employed models and interfaces could be the same or different from the probing process' models and interfaces. When performing an affecting process, the agent may need to use several models at the same time to alter properties that could not be handled by a single model.

In this situation, the agent must perform a composition operation on the existing models to build a new model that could affect its EOI. Essentially, this process determines whether and how certain models could be combined to affect the EOI's properties. The composition operation consists of a parallel operator, a sequence operator[1], and a difference operator that work on multiple levels of model composition. The parallel and sequence operators work on finding ways to combine the models, while the difference operator checks how well the composition process has worked so far.

The agent could own some models and entities while having access to other models and entities through transactions made with other agents. To make this possible, the agent must have currency and suitable transaction models. Within the context of the society proposed here, it is assumed that transaction costs are zero. As a result, agents could gain access to each other's models, making the ownership cost of the models equal to zero unless they are unavailable when being requested. That said, adding transaction costs to the society would put more constraint on learning and make intelligence more dynamical as some agents will create better models and acquire more currency from selling their use to other agents. This would eventually lead to competition in building better probing models, creating diversity in the agents' probing capability. It is not unreasonable to think of a society where some agents develop models to attract other agents' attention. Whatever the situation, each agent uses an administrative process to check whether a model is available, how long it could be used for, how much it would cost, how to carry out a transaction, and how it could be accessed by the agent.

The operators act as an internal interface through which the agent learns over time, in a process called composition search to find out how to compose models to affect its EOIs' properties. Since it is possible that there could be more than one way to compose the models to achieve the result, this creating cognitive redundancy in the search process. This redundancy, however, requires the agent to deal with cognitive complexity arising from the task of finding the right model composition amongst many scenarios, especially with increasing number of models. As a result, complexity makes the composition search's solution at most a local optimum, thus leading to a diversity in

---

[1] Additional operators such as initialization, copying, merging, and terminating could be added to the operation when needed for complex situations.

the composition search's results even when the same search strategy is used. From time to time, the composition operation requires not only internal models, but also external ones acquired through transactions with other entities. The agent is said to be resourceful if it knows how to switch among different approaches to constructing the composition search.

The probing process could be represented by the following incidence matrix.

$$
\begin{array}{c|ccc}
 & e_1 & \cdots & e_l \\
\hline
p_1 & x_{11} & \cdots & x_{1l} \\
\vdots & \cdots & \cdots & \cdots \\
p_k & x_{k1} & \cdots & x_{kl}
\end{array}
$$

Fig. 4. The probing incidence matrix.

Essentially, each agent's probing process could be represented as a weighted hypergraph with it probing models' interfaces p_i as vertices and its EOI e_j as hyperedges (Fig. 4). The properties generated by the probing process are represented by the variable x which changes over time as the agent adjusts his models and the entities' interfaces are updated. x_{ij}, the weight of the vertex p_i is essentially a list of properties of entities e_j learned by the agent through model p_i. In this context, an entity is a hyperedge consisting of weighted vertices that are essentially models used by the agent to construct its representation of the entity. The incidence matrix is thus a random matrix with distribution of its elements capturing the agent's probing process over time.

The agent recognizes an event when there are anticipated changes in its EOIs' property measures. Events could occur internally when the agent adjusts its probing process, causing changes in its assessment of the entities' properties. Events also occur externally when causes of changes come from without. The agent has a goal when it seeks to achieve specific property measures for its EOIs, and goals arise when the agent reacts to changes caused by events.

Intelligence arises when the agent tries to be resourceful in reaching specific goals that are set when it wants to interact with current EOI or new entities it never encountered before. The agent starts with an initial set of goals and proceeds to adjust them over time through two mechanisms. First, it reflects on how certain goals were achieved easily while others were more difficult to realize. Second, it learns how appropriate its goals are through communications with other agents.

Usually, the effecting process requires the composition operation as there could be many models involved in the probing process. This operation could be captured in the following incidence matrix (Fig. 5).

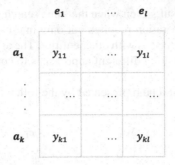

Fig. 5. The affecting incidence matrix.

The matrix's structure is like that of the probing process, but unlike x, y only takes the value of 1 or 0, indicating whether an affecting model a_i is used in the probing operation or not. Like the probing incidence matrix, the affecting matrix only indicates which models were used in the probing process while revealing little as to how the affecting process happened. This could be addressed by the introduction of a diagram wiring scheme and a composition matrix to represent the composition operation (Figs. 6 and 7).

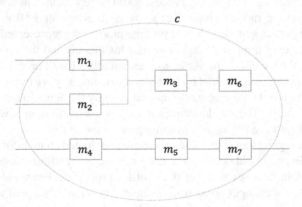

Fig. 6. A composition operation represented as a simple wiring diagram that consists of a parallel operator and a sequence operator working on 7 models. The box represents the model and the wiring the models' input and output. The relationship between the models in this composition operation could be captured by the hypergraph c for two reasons. First, all models would be used in the operation. Second, they could be combined in more than one way to achieve the same result. In this example, the hypergraph also consists of two directed graphs. This simple structure could be expanded to include other data operations such as initialization, termination, copying, merging, and feedback.

Theoretically, this operation could potentially be analyzed by using codesign theory and the concept of compact closed categories that essentially deals with composition

$$m_1 \quad \cdots \quad m_k$$

	m_1	$\cdots$	m_k
m_1	y_{11}	$\cdots$	y_{1l}
$\vdots$	$\cdots$	$\cdots$	$\cdots$
m_k	y_{k1}	$\cdots$	y_{kl}

Fig. 7. A matrix representing to complement the composition hypergraph. In this matrix, m represents the models used in the composition operations, and y takes value of 1, 0 or nothing. When $y_{ij} = 1$, it implies the model m_i is composed sequentially with m_j in this order. When $y_{ij} = y_{ij} = 0$, it implies m_i and m_j are working in parallel in the same or different wiring operations. When y_{ij} has no value, it indicates there is no composition operation working directly on both models.

possibilities [10]. One crucial requirement for this approach to work is that the models' input and output must form preorders that enable the use of enriched categories. Intelligence, however, seems to be able to deal with things that do not seem to have this structure. A workaround would be equipping the agent with a mechanism to transform their models' inputs and outputs into preorders, thus enabling us to analyze the composition process more rigorously. Formally, this could be represented by the following proposition:

$$\bowtie (\mathcal{C}, \mathcal{E}, l) = P(\mathcal{E}) \Leftrightarrow (\mathcal{C}, I, \otimes) is compact closed \, (8)$$

Essentially, (8) states that as soon as the composition operation $\bowtie$ finds a combination of models $\mathcal{C}$ that generate desired properties $\mathcal{P}$ of some EOIs $\mathcal{E}$ under the constraint l, it could be concluded that this process be equivalent to making the symmetric monoidal category $(\mathcal{C}, I, \otimes)$ a compacted closed category. What this proposition implies is that while evolution of intelligence may not be expressed in any precise mathematical formalism, the results of the evolution could be.

This proposition, however, says little about the dynamics of the composition search over time, especially when the agent must perform several operations at once to deal with different EOIs. This issue could be addressed by analyzing the evolution within the framework of hypergraph categories which represent the process as a set of $(\mathcal{C}, I, \otimes)$ nested within and connected to each other via a Frobenius structure [11]. Representing the composition process in the language of hypergraphs and categories would enable appropriate use of algorithms such as the Knuth-Bendix completion rules while dealing with increasing cognitive complexity.

2.2 Learning and Intelligence

Initially, each agent is endowed with a set of EOIs a set of models to probe and affect the EOIs. To better organize knowledge as the number of EOIs increase over time, during the probing process, the agents seek to construct relationships between entities by transforming the hypergraph's incidence matrix into the following adjacency matrix (Fig. 8).

	e_1	$\cdots$	e_l
e_1	r_{11}	$\cdots$	r_{1l}
$\cdot$ $\cdot$ $\cdot$		$\cdots$	$\cdots$	$\cdots$
e_l	r_{l1}	$\cdots$	r_{ll}

Fig. 8. The entity relationship matrix.

The value of r_{ij} is derived from the properties generated by the probing results of two entities e_i and e_j by all models available to the agent (Fig. 4). r is thus a multi-dimensional variable representing what the agents could learn about connections among the EOIs. For example, one value of r_{ij} tells the agent whether two people e_i and e_j come from the same family while another value capture their roles in the family. Looking at these values, the agents could build a family structure that consists of e_i and e_j. In another example, it could learn that both e_i and e_j are cities belonging to a larger entity e_k which is a state or a country.

In general, the agents could construct three kinds of relationship based on the measures captured in the adjacency matrix. The first type enables the agent to group entities into categories according to similarity in their interfaces. In other words, if two entities response to the probing by a similar set of models, they would belong to the same category. The second type looks at relationship within a category in more details by linking together entities that share similar properties. The third type enables the agent to build invariant structural relations that could be applied to different entities, regardless whether they belong to the same category or not, and this could be achieved via exploiting relations among properties. For instance, the agents, by looking at specific properties of e_i and e_j through the probing process, could build a structure to describe their relationship as family members (husband, wife, daughter, dog etc.). they could reuse this structure when encountering a similar context when probing other EOIs, a behavior that has been observed in experiments with mouse brain [12]. While the agents could build structures of relationships among his EOIs, the capability to do so varies as each agent has different probing models with different quality.

Having the capability to find connections and build structural relationships among EOIs, the agents only pay attention to entities likely connected to his current EOI. By doing this, it could learn to anticipate and response quickly and systematically to events. Suppose that there is an event affecting properties of an entity e_i, and the agents know that by using its knowledge of relationships among EOI, e_j and e_k will also be affected.

With this anticipation, they could know how to prepare response to changes in not only e_i but also e_j and e_k.

Even though the agents pay attention to entities related to its EOI, their attention level differs from each other. Some agents care more about specific properties of itself. Other agents start with several interests but over time become slack, while a few keeps increasing their learning capability. That said, when the agents encounter a new entity, they would use all available models to probe it. Results generated by the probing process would give the agents three choices. The first choice is to add the entity to its list of EOIs if there is a hint of a relationship between the entity and some of the existing EOIs. The second option is for the agent to broadcast the encounter to other agents through the administrator. In case some other agents have the right model to deal with the entity, the agent could proceed to make a transaction, if needed. The third choice is to simply ignore it.

When it comes to the composition operation, the agents follow four principles. First, the EOI's input and properties must form or could be transformed into preorders as this enables the agent to establish some sort of hierarchical relationships among the input and the properties. These relationships would in turn enable the composition operation to learn roughly but quickly which combination of inputs or properties will likely be feasible. Second, they do not optimize but only seek heuristically to find approximately right solutions when facing complexity, such as creating changes in an EOI's properties that are within an acceptable limit. This is to avoid the trouble of having to deal with cognitive complexity and to also utilize what the agents know about input, properties and emerging structures learned from reflection.

Third, the agents operate on the model interfaces rather than the actual model constructs. In other words, the agents, assuming that they have full knowledge of the composition search's behaviors, perform simulations of the operation instead trying to use the actual models all the time. By doing this, the agents would avoid the issue caused by unavailability of some models while still operating within a margin of safety. Finally, the composition search could benefit from transactions among the agents, especially when there is a competition among them with the introduction of currency.

To sum up, in the society just constructed, the agents start learning by using available models to probe their EOI through their interfaces. The probing process generates properties associated with the EOI, and these properties enable them to acquire knowledge of the world around them. But probing is not sufficient as they also want to change the properties, and when no existing model could be used, the agents perform a composition operation to create new models from what he has. This composition operation in turn requires an administrative operation to handle communication and transaction involved in model acquisition.

Properties and changes in their measures give rise to events and goals which in turn create a demand for intelligence. In addition, properties enable the agents to build relationships among entities, thus enabling them to learn about these entities not just through knowing individual properties but also through understanding their structural relations. As a result, they understand and respond to events with more complex dynamics. The goal setting capability is thus enhanced accordingly, with intelligence pushed to a new level. In the long term, what matters most to the development of intelligence is the administrative capability to effectively deal with composition search under increasing transaction costs and cognitive complexity.

It is thus sensible to say that one agent is more intelligent than another if the former performs the probing process, the composition search and the administrative function faster and more effectively than the latter does. To achieve these results, the agents use three mechanisms. The first mechanism is a series of snapshots of the learning process he creates when at least one of the following four situations happens. First, the agents create new affecting models through the composition operation. Second, the agents add new entities to his list of EOIs. third, there are changes in the existing entities' properties, and the changes are so significant that they lead to different structure relationships between the entities. Finally, the agents may take a series of snapshots to learn about property relationship which informs whether there is a connection between properties of a specific entity, thus pointing to implications that may benefit the composition operation on the probing models.

The snapshots are essentially entities which would collectively form the agents' memories which could be stored in and retrieved from the following structure. From time to time, an agent could share some of their snapshots to other agents, leading to message exchange, reevaluation of learning capability, and transactions when the snapshots contain models that the other agents need to work on existing and new entities.

$$\left\{ e_i^{mt} \rightarrow \left\{ p^{mt}, x^{mt}, cp^{mt}, a^{mt}, y^{mt}, ca^{mt} \right\} \right\}_{i=1}^{l}, \tag{9}$$

where

$$p^{mt} = \left\{ p_1^{mt}, \ldots, p_k^{mt} \right\}, x^{mt} = \begin{bmatrix} x_{11}^{mt} & \cdots & x_{1k}^{mt} \\ \vdots & \ddots & \vdots \\ x_{k1}^{mt} & \cdots & x_{kk}^{mt} \end{bmatrix}, cp^{mt} = \begin{bmatrix} cp_{11}^{mt} & \cdots & cp_{1k}^{mt} \\ \vdots & \ddots & \vdots \\ cp_{k1}^{mt} & \cdots & cp_{kk}^{mt} \end{bmatrix} \tag{10}$$

$$a^{mt} = \left\{ a_1^{mt}, \ldots, a_l^{mt} \right\}, y^{mt} = \begin{bmatrix} y_{11}^{mt} & \cdots & y_{1l}^{mt} \\ \vdots & \ddots & \vdots \\ y_{l1}^{mt} & \cdots & y_{ll}^{mt} \end{bmatrix}, ca^{mt} = \begin{bmatrix} ca_{11}^{mt} & \cdots & ca_{1l}^{mt} \\ \vdots & \ddots & \vdots \\ ca_{l1}^{mt} & \cdots & ca_{ll}^{mt} \end{bmatrix} \tag{11}$$

Essentially, an agent m's snapshot t is a list in which element i describes an association between the entity e_i and the corresponding probing models (p), property matrix (x), composition matrix for the probing models (cp), affecting models (a), affecting matrix (y), and composition matrix (ca) used by m in the probing and affecting process. The snapshots are a rough representation of what happened in the past since the agent only records them in certain situations, thus making a trade-off between perfect memory and computational complexity.

The second mechanism is credit assignment which essentially ranks composition searches by reflecting on past snapshots. When probing a new EOI, the agents would favor probing models and composition solutions that have higher credit assignment. One simple strategy is to extract and give higher credit to models that remain invariant in the probing and composition operations over the reflection period. If these models turn out to be effective, they will in turn reinforce their credit ranking. This circle will allow the agents to reduce search time and avoid unnecessary computational complexity when probing new EOIs or performing composition operation. The strategy, however, may put them in a difficult situation when they overuse the credit assignment without considering the context applied to. Regardless of the strategy used, the credit assignment essentially deals with estimating the following probability

$$P(\bowtie (C, \mathcal{E}, l, \mathcal{P}(\mathcal{E}))| \bowtie (C^*, \mathcal{E}^*, l^*, \mathcal{P}(\mathcal{E}^*))), \quad (12)$$

where * denotes a degree of similarity between elements of other composition operations and that of the current operation.

The third mechanism is parallel processing of the probing operation, the administrative function and the credit assignment process. The idea is rather simple. Doing things in parallel, if the outcomes could be combined in the right way, would speed up the computational process significantly, especially when the number of tasks increases. Performing the composition search concurrently, however, may be more difficult unless the agents have enough redundancy in its architecture to support the operation. The reason is that as some models are used during one instance of composition search, they will not be available for use in another instance unless the agents have some copies ready for the operation. This architecture issue is discussed in the next section.

2.3 Network Structure

Having developed the constructs for the learning process and the mechanisms to create memories that enable agents to be more intelligent, they need a place to store the constructs and operate the mechanisms. What would the architecture of such place look like? The first feature of the architecture would be a network-like structure, since the learning process relies on connections between the probing, affecting and administrative operations. Organizing these operations around networks would enable them to be carried out in parallel efficiently.

The second feature would be a separation of the architecture into two parts in term of operational speed and complexity. The first part consists of computing nodes focusing on coordinating probing, affecting and administrative tasks in parallel. The nodes do not carry out the operations but perform a supervision role instead. This part also contains nodes linked to the actual models used in the operation. They serve as a temporary memory that keeps information about the EOIs and model interfaces. This enables the supervisory functions to work on these models and entities at rapid pace and, when possible, in parallel fashion.

This process works with the principle that knowing what comes in and what comes out through the interface would be reasonably enough to make judgements as to whether the probing, affecting and administrating operations would work or not. Besides, the nodes always make the interface available when needed by these operations, while actual models may not. Doing so, the agent is sacrificing uncertainty in access and slow precision offered by operations on actual models for high availability and rapid approximation delivered by implementation on the interfaces.

The second part is where snapshots and the actual models' constructs are kept. The purpose of keeping these entities in a separate place is twofold. First, it will over time require more and more permanent memory capacity and putting these in the first part of the architecture will create a server constraint temporal memory needed for crucial operations. Second, working directly on the models and wait for actual results is much slower than working on their interfaces with the presumption that the interfaces would do what the actual models would do, especially when many models get involved in the operations. Finally, it is less computationally complex to have a verification on actual model at the end of an operation rather than at each step in the process. That said, some models that are used frequently and have simple constructs could still be kept in the first part, especially those serving as elementary constructs upon which other models are built.

Next, what languages do the operations use to communicate and coordinate? For the first part of the architecture, it must be simple enough to enable the operations to be carried out rapidly in parallel without causing computational overhead. A binary language appears to be a sensible solution since the elements of this language are just bits. And the rules for this language should be based on basic logical and bitwise operations. Thus, at the center of the architecture are binary networks that perform the probing, affecting, and administrating operations. Recent developments in studies of binary neural networks pint to the potential of this computational model [13]. The next challenge is to construct appropriate coding for the network. One could imagine watching such networks at work when some nodes become active, they represent models that are being used in the operations. And when they become active together, they are used in a composition search, and we could say that when they fire together, they wire together. These nodes are not models that perform complicated computations, instead they serve as an interface to those models.

For the constructs stored in the second part, they could be represented in any languages that satisfy three criteria. First, the language must have an interface to the binary language used by the first part, thus enabling communications between the actual models and the representative interfaces. Second, the language must support execution of parallel operations on the constructs initiated by the binary network. Especially, it must enable parallel operations to be carried out without causing unnecessary computational overhead. Finally, it must has a common interface to enable the agents to exchange messages and make transaction with each other, and these tasks could be performed in a parallel implementation made possible by development of mature and robust message-based concurrent programming language such as Erlang [14].

Knowing how these constructs should be implemented and what performance would be expected from such implementation would allow validation of the architecture presented here. What this paper hopes to accomplish is to present new conceptual elements as core design artifacts of intelligence. Being a proposal of new ideas, the paper has not addressed issues such as how credit assignments evolve and how the agents deal with cognitive complexity during the composition search process over time. We believe that the best avenue to seek answers for these questions is in actual implementation of the architecture, and this our goal in the next endeavor.

3   Conclusion

This paper approached the study of intelligence from a network computation perspective based on abstract structures that could be applied to different contexts rather than just specific tasks. These structures were constructed by observing emerging properties of intelligence that essentially deals with cognitive redundancy and complexity. Using the concept of agents, the study is also concerned with social aspect of intelligence. As a result, agents would be expected to exhibit administrative behaviors sooner or later in the evolution of its intelligence as they learn how to deal with cognitive redundancy and complexity effectively. From a design perspective, the basis to support these behaviors is a network architecture amenable to structural analysis based on network theories and hypergraph categories. Key conceptual constructs of this architecture were developed and discussed with the purpose of creating a sound basis for further investigations in future studies.

References

1. Simon, H.A.: The Sciences of the Artificial. MIT Press, Cambridge, Mass (1996)
2. Minsky, M.: The Emotion Machine: Commensense Thinking, Artificial Intelligence, and the Future of the Human Mind. Simon & Schuster, New York (2006)
3. Berge, C.: Hypergraphs: Combinatorics of Finite Sets. Elsevier (1984)
4. Baader, F., Nipkow, T.: Term Rewriting and all that. Cambridge University Press (2012). https://doi.org/10.1017/CBO9781139172752
5. Gorard, J.: Some quantum mechanical properties of the wolfram model. Complex Syst. **29**(2), 537–598 (2020)
6. Zador, A.M.: A critique of pure learning and what artificial neural networks can learn from animal brains. Nat. Commun. **10**(1), 1–7 (2019)
7. Simon, H.A.: Administrative Behavior: a Study of Decision-Making Processes in Administrative Organizations. Free Press, New York (1997)
8. Knuth, D.E., Bendix, P.B.: Simple word problems in universal algebras, pp. 342–376. Springer, Automation of Reasoning (1983)
9. Nicolau, D.V., et al.: Parallel computation with molecular-motor-propelled agents in nanofabricated networks. Proc. Natl. Acad. Sci. **113**(10), 2591–2596 (2016)
10. Censi, A.: A mathematical theory of co-design. arXiv:151208055 (2015)
11. Fong, B., Spivak, D.I.: Hypergraph categories. J. Pure Appl. Algebra **223**(11), 4746–4777 (2019)
12. Whittington, J.C.R., et al.: The Tolman-Eichenbaum machine: unifying space and relational memory through generalization in the hippocampal formation. Cell **183**(5), 1249–1263 (2020)
13. Valencia, R., Sham, C.W., Sinnen, O.: Using Neuroevolved Binary Neural Networks to solve reinforcement learning environments. In: 2019 IEEE Asia Pacific Conference on Circuits and Systems (APCCAS), pp. 301–304. IEEE (2019)
14. Armstrong, J.: Making reliable distributed systems in the presence of software errors (Doctoral dissertation) (2013)

Taking Cognition Seriously
A Generalised Physics of Cognition

Sophie Alyx Taylor[1]([✉]) [iD], Son Cao Tran[1] [iD], and Dan V. Nicolau Jr.[1,2,3] [iD]

[1] School of Mathematics, Queensland University of Technology, Brisbane, Australia
`sophie.taylor@hdr.qut.edu.au`
[2] Respiratory Medicine Unit, Nuffield Department of Medicine, University of Oxford, Oxford, UK
[3] Centre for Clinical Research, University of Queensland, Brisbane, Australia

Abstract. The study of complex systems through the lens of category theory consistently proves to be a powerful approach. We propose that cognition deserves the same category-theoretic treatment. We show that by considering a highly-compact cognitive system, there are fundamental physical trade-offs resulting in a utility problem. We then examine how to do this systematically, and propose some requirements for "cognitive categories", before investigating the phenomenona of topological defects in gauge fields over conceptual spaces.

Keywords: Mathematics of cognition · Topological field theories · Applied category theory · Computational cognitive architecture · Computation in the Schwarzschild metric

1 Introduction

The study of thought has occupied academia for millennia, from philosophy, to medicine, and everything in between. We demonstrate a promising approach to take advantage of the powerful tools developed by other disciplines in our study of cognition, by emphasising the role that category theory – a foundation of mathematics – ought to play.

We first motivate our approach by considering the effects of general relativity on a highly compact cognitive agent, and noting the structural similarities between metric field theories in physics, and the dynamics of mental content in conceptual spaces. We feel that these similarities practically beg for the study of field theories in *conceptual spaces themselves*, and consider the topological defects that arise; interpreting them, as is done in physics, as *particles of thought*.

2 Background

We first recall some required background which we build upon.

© ICST Institute for Computer Sciences, Social Informatics and Telecommunications Engineering 2021
Published by Springer Nature Switzerland AG 2021. All Rights Reserved
T. Nakano (Ed.): BICT 2021, LNICST 403, pp. 230–243, 2021.
https://doi.org/10.1007/978-3-030-92163-7_19

2.1 Cognitive Architecture and Conceptual Spaces

To model a dynamical system, we require specification of two things: How the data is to be represented, and the system that operates on it. In our approach, we use conceptual spaces and cognitive architectures, respectively.

Conceptual Spaces. *Conceptual spaces* [7,22] provide a convenient middle-ground approach to knowledge representation, situated between the symbolic and associationist approaches in terms of abstraction and sparseness. A conceptual space is a tensor product of conceptual *domains*, which can be arbitrary topological spaces equipped with certain measures. They provide a low-dimensional alternative to vector space embeddings, which often have no clear interpretation of their topological properties [3]. Conceptual spaces represent concepts as regions in a topological space. Because (meta-)properties of the conceptual spaces depend on context, this suggests that a proper description involves the notion of a (pre)sheaf. Conveniently, they also provide a well-motivated, natural construction for prior probabilities based on the specific symmetries of the domain [5].

Cognitive Architecture. Cognitive architectures are an system-level integration of many different aspects of "intelligence". It can be useful to think of cognitive architectures as the cognitive analogue of a circuit diagram; specifying the hardware/operating system (in the case of robotic agents) or the wetware (in the case of biological agents) that the processes deliberate thought, intentions and reasoned behaviour occur. They generally take a "mechanism, not policy" approach: just like a mathematician has the same brain structure as a carpenter, a given cognitive architecture can produce wide range of agents, depending on what has been learned.

Cognitive architecture has been studied in the context of Artificial Intelligence for at least fifty years, with several hundred architectures described in the literature. Laird [12] describes the SOAR cognitive architecture, one of the most well-established examples in the field. Several surveys have been presented, such as Goertzel, Pennachin, and Geisweiller [8] and Kotseruba and Tsotsos [11]. Huntsberger [9] describes one particular architecture designed for human-machine interaction in space exploration.

Generalised Problem Space Conceptual Model. As a concrete example of a cognitive architecture, we show a modified version of the Problem Space Conceptual Model [12] in Fig. 1. It is reminiscent of a multi-head, multi-tape Turing machine; but with transition rules stored in memory, too. The core idea of the PSCM is that operators are selected on the basis of the contents of working memory, which is essentially short-term memory. Production rules elaborate the contents of working memory until quiescence; operators are essentially in-band signals about which production rules should be fired. If progress can't be made, an *impasse* is generated; from here, we can enter a new *problem space*. This is essentially a formalisation of a "point of view" to solve a problem.

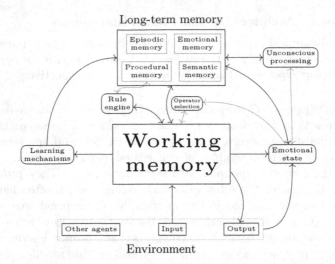

Fig. 1. A slight generalisation of the Problem Space Conceptual Model. Control lines are drawn in orange, while data flow is depicted in blue. Rounded corners on a black indicate that it while it has fixed functionality, it exposes control knobs to working memory, just like Direct Memory Access in a computer. (Color figure online)

Realisation of Cognitive Architecture. Cognitive architectures are *abstract models*, which need to be *concretely realised*. We almost always study cognition through a realisation via neuroscience, which itself is realised via biology, then chemistry, then physics. This is depicted in Fig. 2. Alternatively, we may realise it in the form of robotics and software simulations; like with neuroscience, these are also built upon a tower of realisations.

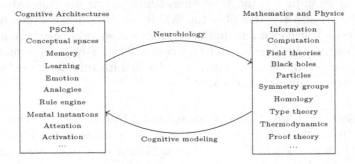

Fig. 2. The abstract models of cognitive architectures are typically concretely realised by way of neurobiology.

2.2 Limits on Computation

Computation has its limits; most famously, the halting problem. We can find limits of computation in at least two general classes: limits imposed by computation in-and-of-itself, and limits imposed by physics on the embodiment of a computation.

Busy-Beaver Machines. Limits on maximum output of a Turing machine for a given number of states, and number of transitions, $BB(n)$. The Busy Beaver game is to construct an n-state Turing machine which outputs the largest amount of "1"s possible, *then halts*. A variation on this is to count the number of steps it takes for the said Busy Beaver machine to halt. If you look at this as a function in n, whether counting the size of the output or the transitions, you have an extraordinarily fast-growing function, quickly outpacing Ackermann's function. You can use this to decide whether n-state Turing machines halt, simply by waiting BB(n) steps. If it takes longer than that, then you know it will never halt, by definition, because it would be the new BB(n). Of course, BB(n) itself is uncomputable in general: We have no free lunch.

The Bekenstein Bound. The Bekenstein bound is a limit on information density:

$$I \le \frac{2\pi RE}{\hbar c \ln 2} \quad \text{bit} \tag{1}$$

where R is the radius of the system, E is the energy of the system, $\hbar$ is the reduced Planck's constant, and c the speed of light [21]. If one assumes the laws of thermodynamics, then all of general relativity is derivable from Eq. 1 [10]. If you fixes the radius R, then as you try to fit more and more information in the given volume, you must eventually start to increase the energy in that volume. Eventually, the Schwarzchild radius of that energy equals R, and you have a black hole. This occurs when Eq. 1 is an equality.

The Margolus-Levitin Bound. The Margolus-Levitin bound is a bound on the processing rate of a system. A (classical) computation must take at least $\frac{h}{4E}$. To decrease the minimum time to take a computational step, you must increase the energy involved. If this energy is contained in a finite volume, then you suffer the same problem arising from the Bekenstein bound; eventually the Schwarzchild radius catches up to you. It's important to note, however, that this assumes no access to quantum memory.

2.3 Gauge Fields and Their Symmetries

A *gauge field* is an approach to formalising what we mean by a *physical field*. We think of a gauge field as data "sitting above" the points in a space (a *principle bundle*), as shown in Fig. 3, combined with a transformation rule (a *connection*) to ensure the data transforms sensibly with a change of test point.[2] Gauge fields typically have a very useful property: We can reparameterise them to be more convenient to work with. For example, we can change the electric and magnetic potentials in an electromagnetic field by a consistent transformation derived from arbitrary data, yet the exact same physical effects will occur. This is called *gauge fixing*.

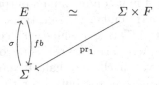

Fig. 3. A gauge field F over a spacetime Σ.

2.4   Topological Defects

Topological defects are an important notion in a number of scientific and engineering fields. Intuitively speaking, they are *imperfections* in what we call an *ordered medium*; roughly, something which has a consistent rule for assigning values to a point in space, like a gauge field [15]. Some familiar examples include:

- Grain boundaries in metals, consisting of disruptions in the regular lattice placement of the constituent atoms,
- Singularities and vortices in vector fields, such as black holes and tornadoes, respectively, and
- Gimbal lock is a topological defect in the Lie algebra bundle $\mathfrak{spin}(3) \to T^3$ of rotations over the Euler angles.

In general, defects arise when there are non-trivial homotopy groups in the principle bundle of the gauge field; that is, when you can make a loop in the bundle space, but you can't shrink it down to a single point.

These are non-perturbative phenomena, arising due to the very *nature* of these systems. These manifest themselves as *(psuedo-)particles* or higher branes, such as membranes. However, due to the risk of confusion between the homonyms "brain" and "brane", we will call them all particles, regardless of their dimension.

Dynamics. What makes defects particularly interesting is that they can evolve over time. Dynamical behaviour in the base space can lead to dynamical behaviour of defects; for example, the movement of holes in a semiconductor, or the orbit of a small black hole around a larger one.

Temperature and Spontaneous Symmetry Breaking. Another common phenomenon is that there is a *temperature* associated with the defect, which controls both its dynamics and its very existence. The obvious example of a temperature parameter is thermal temperature; if you melt a bar of iron, you won't have any more crystal structure to be defective. A more challenging one is temperature in superconductors or superfluids; raise it too high and Cooper pairing no longer works, so you lose superconductivity and superfluidity, respectively. These drastic changes in behaviour are *phase changes*.

Consider the opposite situation: Freezing a liquid to a solid. If you lower the temperature to a point where defects start appearing, then those defects have to actually appear *somewhere*. The configuration of the defects will be just one possible configuration out of a potentially astronomically large space of possibilities (such as the precise atoms involved in a grain boundary), chosen at random. This phenomena is called *spontaneous symmetry breaking*, and is responsible for the masses of elementary particles, via the Higgs mechanism.

2.5   Category Theory and Its Slogans

Category theory is a foundation of mathematics, as an alternative to set theories like ZFC. Instead of focusing on membership, category theory focuses on *transformation* and *compositionality*. Its methodology for studying an object is to look at how that object relates to other objects. There are many excellent introductions to category theory aimed at different audiences (such as [6,14,16,20]). These introduce the basic notions and show the versatility of category theory in a wide range of fields.

Categories and Functors. A category is a collection of objects, and a collection of morphisms between those objects. For example, the category of sets has sets as objects, and functions as morphisms. Different classes of category allow different constructions; for example, monoidal categories allow *pairing* of objects. You can translate between categories with *functors*, which are themselves just morphisms in the category of categories.

Adjunctions. A common situation in mathematics arises when you want to translate back and forth between two types of structures via a pair of functors, but the structures aren't isomorphic. The best we can do is approximate an isomorphism; this is called an *adjunction* [16].

Enrichment. *Enriched categories* are a generalisation of plain categories. Whereas in standard category theory, the morphisms between objects are described by their homset, a category enriched in a monoidal category $\mathcal{V}$ has hom-objects from $\mathbf{Ob}(\mathcal{V})$ [6, pp. 56,139]. Ordinary categories are just categories enriched over the category of sets [6, p. 87].

An Approach to Science and Mathematics. Let us take a step back for a moment, and consider *how* science should be done. Bill Lawvere offers this:

> When the main contradictions of a thing have been found, the scientific procedure is to summarize them in slogans which one then constantly uses as an ideological weapon for the further development and transformation of the thing [13].

Let us take this approach. What are some slogans for category theory?

- Better to have a nice category of objects than a category of nice objects [4]
- Dialectical contradictions are adjunctions [13]
- Find and characterise the principal adjunction of any problem

We shall now provide some motivation for our work.

3   Motivation

We motivate our approach by considering arbitrary cognitive systems in terms of physics. First, we consider the consequences of a physically dense cognitive agent; second, we consider the role self-parameterisation in conceptual spaces.

3.1   The High-Density Regime of Cognition and Its Consequences

Let us consider the high-density regime of cognition; that is, where the cognitive agent's radius is close to its Schwarzchild radius. We shall assume that the agent can learn from its experiences. As the agent experiences the world, it may learn new information,[1] which requires storage in the agent's memory. Consider this increase in information in the context of Eq. 1: more and more energy will be required to store it in a given volume. This, in turn, results in an increase in the Schwarzchild radius. As we are already close to our Schwarzchild radius, and we presumably don't want to collapse into a black hole,[2] we must increase our containment radius; this results in both the information storage *and* processing gadgets spreading out. This results in the average distance between two pieces of information in the containment volume increasing, which in turn results in increased average propagation delay to shunt information around. Thus, the average serial processing rate will decrease—in other words, *learning can make you slower*. This isn't even taking into account gravitational time dilation, which will only further enhance the slowdown relative to an outside observer, as well as the energy required to interact with the external environment.

---

[1] These experiences may allow generalisation of existing information and thus discarding of the individual cases; but over time, there will be a point where generalisations can't be reasonably made.

[2] It is ill-advised to become a black hole, if only to be able to affect the external environment in a reasonable manner (but see [1]).

Setting. We consider an agent comprised of an incompressible fluid which is spherically symmetric and non-rotating, and electrically neutral, resulting in the *interior Schwarzschild metric* [19]. In Schwarzschild coordinates, the line element is given by

$$
c^2 d\tau^2 = \frac{1}{4} \left(3\sqrt{1 - \frac{R_S}{R_g}} - \sqrt{1 - \frac{r^2 R_S}{R_g^3}} \right)^2 c^2 dt^2
$$
$$
- \left(1 - \frac{r^2 R_S}{R_g^3} \right)^{-1} dr^2 - r^2 \left(d\theta^2 + \sin^2 \theta \, d\varphi^2 \right),
$$

(2)

where

(t, r, θ, ϕ) = The spacetime coordinates in $(+ - --)$ convention
R_S = The Schwarzschild radius of the agent
R_g = The value of r at the boundary of the agent, measured from the interior.

Slowdown Due to Propagation Delay. If we consider an instantaneous path through our agent – that is, one where $dt = 0$ – we can see from Eq. 2 that when R_g is close to R_S, the radial term blows up as the path approaches the surface; indeed, we can see that the distance required to be travelled to travel to the surface when $R_g = R_S$ is infinite.

Slowdown Due to Gravitational Time Dilation. The time dilation experienced by the agent compared to an observer at infinity increases with density. The scaling compared to the outside observer is given by

$$
\sqrt{1 - \frac{R_S}{r}}
$$

(3)

valid for $r > R_S$.

Extra Energy Required to Interact with the External Environment. Not only will the extreme gravitational field reduce rate of processing, it will also increase the energy required to interact with the external environment. Consider the case of our agent trying to communicate with a distant observer via some protocol based on the exchange of radio signals. If there is any specification of frequency of the carrier signal, then we must account for the gravitational redshifting of our agent's emissions. This redshifting results in longer wavelength signals; equivalently, lower frequency signals. Because the energy of the photons comprising the signal is directly proportional to its frequency, this can be stated in terms of *lower energy signals*. Thus, in order to ensure outbound transmissions meet the frequency specifications of the protocol, the photons must be emitted with extra energy to overcome the extreme curvature of the gravitational field.

In the simplest case, assume the exterior Schwarzschild metric [18]. To signal an outside observer at infinity with a photon of a given energy E emitted at a radius R_E, the photon will have to be emitted with an energy increased by a factor of

$$\frac{R_E}{\sqrt{R_E(R_E - R_S)}} - 1. \tag{4}$$

Some Tasks Are Too Complex to Be Solved in a Given Volume. We have thus seen that there is a *fundamental physical trade-off* between learning, and the time it takes to solve an arbitrary task. This immediately implies a number of things:

- There is a trade-off between the general capability of an agent, and its average reactivity,
- Some complex tasks with a time requirement are simply too complicated for any agent to solve the task in a given volume of space,
- As you increase the average capability of a group of agents, less of them will be able to fit in a given volume.

Lattice of Cognitive Skills. We can use these limits to create various partial orders:

- We can rank tasks by the minimum and maximum volume of agents capable of solving it,
- We can rank agents by their capabilities with respect to physically-embodied Busy-Beaver machines,
- We can rank space-time volumes by combinations of the above

3.2   Self-parameterisation of Behaviour

Let us consider our second motivational example: the self-parameterisation of behaviour. The behaviour of an agent depends on its accumulated knowledge up until that point, in particular, its procedural knowledge. Its knowledge, in turn, depends on its past behaviour acquiring the knowledge. This self-referential quality appears in physics, in the form of *metric field theories*. Consider, again, general relativity: The metric tensor controls the dynamics of a system, and the distribution of that system determines the metric tensor. The previous example considered the *effects* of general relativity on a cognitive agent, but the agent didn't particularly play an *active* role in our considerations.

But, why consider only general relativity? Can we consider other metric field theories? For that matter, do we even have to consider *physical* field theories?

4   Goal

Consider the position our motivational examples has put us in: Instead of considering cognition as an embodied agent comprised of wetware or hardware, we just re-enacted the old "assume a spherical cow in a vacuum" joke. We didn't bother trying to determine *how* a dense sphere of incompressible fluid could embody a cognitive agent; it turned out we didn't even need to! We just assumed that the translation from an abstract model of cognition, to the concrete sphere of cosmic horror, and then back again to an abstract model preserved the semantics of cognition, whatever they may be.

4.1   Taking Advantage of Tools from Other Disciplines

Let us now state what we wish to do, in order to find a solution. We wish to use the mathematical tools from other disciplines in order to study cognition. How is this typically done?

The Physics of Cognition, by Way of Biology. As shown in Fig. 2, we translate through a chain of "domains" of science before reaching physics. Each translation introduces an extraordinary amount of complexities that exist solely due to choosing a specific concrete realisation *at each stage*. While, of course, neurology and biology are *incredibly* important subjects of study in humans – if only for their phenomenological observations, not to mention the pathophysiological importance – they are complicating factors in the study of the phenomena of cognition in-and-of-itself.

We Only Need to Find Nicer Realisation Morphisms Which Preserve Behaviour. Instead of considering the familiar realisation via neurobiology, the realisation transformation only needs to preserve the structures and behaviours of interest; that is, *we only need to find toy models*. To do this, we need to find a nice class of "cognitive categories", and adjunctions out of them, as shown in Fig. 4. A slightly more category-theoretic diagram is shown in Fig. 5.

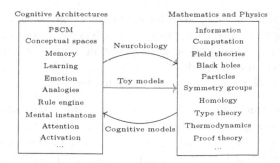

Fig. 4. All we need to do is to preserve the behaviours, not any particular concrete realisation strategy.

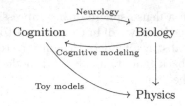

Fig. 5. Transformation expressed in a more traditional categorical diagram

4.2 Cognitive Categories

If we are going to use category theory to study cognition, then we ought to specify what cognitive categories actually are. This is an open problem, but some plausible requirements include:

- Subcategories of cognitive categories ought to include Turing categories, in order to capture computational behaviour.
- There should be an opportunity for enrichment in a category of conceptual spaces.

5 Example: Topological Defects in Conceptual Spaces

Let us consider the metric field analogy a little deeper. Since many conceptual spaces of interest have a genuine geometric structure, and we can form fibre bundles over them which allow parallel transport, then we can at least *consider* gauge fields.

5.1 Cognitive Gauge Fields

There are a lot of potential options for cognitive gauge fields. Some are generic, which might apply to any conceptual space; whereas some might only apply to certain classes of conceptual space. We assume some mechanism for smooth interpolation in cases of noisy discrete data.

Some example plausible generic gauge fields include:

- The activation value of a specific memory is related to its probability of being recalled due to a query. The higher the activation value, the more it will be recalled. These values can spread to adjacent memories. Many "forgetting" mechanisms are based on forgetting memories with low activation value.
- We can consider the emotional state of an agent when storing the memory, decomposed into a set of *valuation* and *valence* affect dimensions.
- We can also consider the subjective importance of a memory, which is separate to its activation; an example in procedural memory is a rule saying to not completely flood a room with water if there are people in it. It's not very *likely* to be subject to recall, but it's certainly an important rule!

Some plausible non-general gauge fields might be:

- Trustworthiness of data gathered socially.
- Difficulty of tasks and behaviours. This can evolve based on experience and better understanding of the situation.

5.2 Defects and Their Dynamics: Particles of Thought

Some cognitive gauge fields will have non-trivial topology, resulting in topological defects. As the underlying conceptual spaces evolve, whether contents or the topology itself, we might observe dynamical evolution of these defects. Thus, we might (not-so-metaphorically) call these "particles of thought".

Leaving aside whether such a thing has a meaningful interpretation,[3] we can at least ask more about the nature of such things.

5.3 Production Rules as Potentials

Can we encode production rules as a something akin to a potential? If so, what actually generates those potentials?

5.4 Transmission of Influence and Cognitive Gauge Bosons

How is the influence of a gauge field propagated? Is it wave-like, as in classical physics, or are there 'force-carrying particles', like gauge bosons in particle physics?

5.5 Phase Changes

Are there phase changes? That is, is there some order parameter where the particles are only manifest in a given range of parameter values? Does this relate to switching problem spaces?

5.6 Cognitive Event Horizons

Are there "cognitive event horizons", where there are boundaries from which the effects of a particle can never escape, not just effects on its surrounding particles or memories, but on behaviour too? If so, how do they form? How do they evolve? Are there mechanisms to affect the underlying topological structure of a conceptual space? Is there a "maximum resolution" to some conceptual spaces as a result, analogous to the Schwarzschild radius in general relativity?

---

[3] This almost certainly depends on the conceptual space and gauge field involved.

6   Discussion

There is an important consideration to be made when talking about theoretical modelling: we must stress when we are only talking about *effective theories*; that is, theories which model the *effects* of cognition, but do not make any claims as to whether there is any actual *causal* connection with the reality of cognition. The ontological status of particles of thought certainly rests on the status of conceptual spaces, and gauge fields over them. Further, assuming they *are* ontologically valid, whether they actually have any causal role requires considerable further study.

6.1   Future Research

We have a number of interesting questions prompted by our approach:

- How might we study the flow of information throughout an agent's lifetime? Can this be linked to "particles of thought"?
- How can an agent perceive the Self? Does an agent who is aware of itself encounter, for example, the Barber's paradox? How can it reason about things which are not true? What are the connections with paraconsistent logic?
- How can we more fruitfully take advantage of topological data analysis?
- How does analogy work in our approach? Does it rely on the homology structure of the relevant conceptual spaces having particular forms?
- What is a good model for various learning mechanisms; both of mental content itself, and of new conceptual spaces?
- What group structures over different conceptual spaces can we find to yield different field theories?
- Behaviour and cognition depends heavily on emotional state [17]. What is the most appropriate conceptual space to represent this, and do they have any psuedoparticles?

References

1. Andrews, G.: Black hole as a model of computation. Results Phys. **13**, 102–188 (2019). https://doi.org/10.1016/j.rinp.2019.102188
2. Baez, J.C., Muniain, J.P.: Gauge Fields. Knots and Gravity. World Scientific Publishing Company, Singapore (1994)
3. Bolt, J., Coecke, B., Genovese, F., Lewis, M., Marsden, D., Piedeleu, R.: Interacting Conceptual Spaces I: Grammatical Composition of Concepts (2017). arXiv: 1703.08314, 29 September 2017
4. Corfield, D.: Modal Homotopy Type Theory: The Prospect of a New Logic for Philosophy. University Press, Oxford (2020)
5. Decock, L., Douven, I., Sznajder, M.: A geometric principle of indifference. J. Appl. Logic **19**, 54–70 (2016)
6. Fong, B., Spivak, D.I.: An Invitation to Applied Category Theory: Seven Sketches in Compositionality. Cambridge University Press, Cambridge (2019). https://doi.org/10.1017/9781108668804

7. Gärdenfors, P.: Conceptual Spaces: The Geometry of Thought. MIT Press, Cambridge (2004)
8. Goertzel, B., Pennachin, C., Geisweiller, N.: Brief survey of cognitive architectures. In: Engineering General Intelligence, Part 1. ATM, vol. 5, pp. 101–142. Atlantis Press, Paris (2014). https://doi.org/10.2991/978-94-6239-027-0_6
9. Huntsberger, T.: Cognitive architecture for mixed human-machine team interactions for space exploration. In: 2011 Aerospace Conference, pp. 1–11 (2011). 10/cq4m28
10. Jacobson, T.: Thermodynamics of spacetime: the Einstein equation of state. Phys. Rev. Lett. **75**(7), 1260–1263 (1995). https://doi.org/10.1103/PhysRevLett.75.1260. arXiv: gr-qc/9504004
11. Kotseruba, I., Tsotsos, J.K.: A Review of 40 Years of Cognitive Architecture Research: Core Cognitive Abilities and Practical Applications (2016). arXiv: 1610.08602, 26 October 2016
12. Laird, J.E.: The Soar Cognitive Architecture. MIT Press, Cambridge (2012)
13. Lawvere, F.W.: Categories of space and of quantity. In: Echeverria, J., Ibarra, A., Mormann de Gruyter, T., (eds.) The Space of Mathematics. Boston, Berlin (1992). https://doi.org/10.1515/9783110870299.14
14. Lawvere, F.W., Schanuel, S.H.: Conceptual Mathematics: A First Introduction to Categories. Cambridge University Press, Cambridge (2012)
15. Mermin, N.D.: The topological theory of defects in ordered media. Rev. Mod. Phys. **51**, 591–648 (1979). https://doi.org/10.1103/RevModPhys.51.591
16. Riehl, E.: Category Theory in Context. Dover, New York (2016)
17. Rosales, J.-H., Rodríguez, L.-F., Ramos, F.: A general theoretical framework for the design of artificial emotion systems in autonomous agents. Cogn. Syst. Res. **58**, 324–341 (2019). 10/ggh7sr
18. Schwarzschild, K.: On the gravitational field of a mass point according to Einstein gleghber theory. Sitzungsber. Preuss. Akad. Wiss. Berlin (Math. Phys.) **1916**, 189–196 (1916). arXiv: physics/9905030
19. Schwarzschild, K.: On the gravitational field of a sphere of incompressible fluid according to Einstein's theory. Sitzungsber. Preuss. Akad. Wiss. Berlin (Math. Phys.) **1916**, 424–434 (1916). arXiv: physics/9912033
20. Spivak, D.I.: Category Theory for the Sciences. MIT Press, Cambridge (2014)
21. Tipler, F.J.: The structure of the world from pure numbers. Rep. Prog. Phys. **68**(4), 897–964 (2005). https://doi.org/10.1088/0034-4885/68/4/R04. arXiv: 0704.3276
22. Zenker, F., Gärdenfors, P. (eds.): Applications of Conceptual Spaces. SL, vol. 359. Springer, Cham (2015). https://doi.org/10.1007/978-3-319-15021-5

Accelerated Virus Spread Driven by Randomness in Human Behavior

Huber Nieto-Chaupis[✉]

Universidad Autónoma del Perú,
Panamericana Sur Km. 16.3 Villa el Salvador, Lima, Peru

Abstract. In this paper is demonstrated that the morphology of infection's curve is a consequence of the entropic behavior of macro-systems that are entirely dependent on the nonlinearity of social dynamics. Thus in the ongoing pandemic the so-called curve of cases would acquire an exponential morphology as consequence of the human mobility and the intensity of randomness that it exhibits still under social distancing and other types of social protection adopted in most countries along the first wave of spreading of Covid-19.

Keywords: Epidemiology · Global pandemic · Human behavior

1 Introduction

From the end of 2019, the so-called Corona virus strain [1] have spread along the globe yielding more than 100 millions of infections and more than 3 millions of fatalities. To date this paper is written it is fully unknown if more strains are still to be expected despite of the fact that vaccine programs have been done in most countries.

As noted in [2], it was an expectant view about the arrival of a strong pandemic with similar characteristics as the AH1N1 seen in the outbreak of 2009. Certainly the ongoing situation of pandemic cannot be comparable to that scenario at the past decade [3]. Clearly the ongoing scenario has surpassed expectant models, fact that leads to establish that the corona virus is fully adaptable to various conditions [4] such as for example climate and altitude. In this manner, the virus has achieved to penetrate all geographical places in all different physiological hosts. The biochemical adaptability of virus is seen in the highest rates of infections in all continents. On the other hand one can wonder about the role of humans to extend the infections in all places. Thus one can anticipate at the human behavior and the nonlinear components. In this way emerges various questions about the factors that are favorable to the propagation of strain and the causes and the interralation among them that would be relevant to explain the fast increasing of infections per unit of time [5].

In this paper, the human behavior characterized by being one of random character and also seen as a factor of infection, is mathematically modeled through

© ICST Institute for Computer Sciences, Social Informatics and Telecommunications Engineering 2021
Published by Springer Nature Switzerland AG 2021. All Rights Reserved
T. Nakano (Ed.): BICT 2021, LNICST 403, pp. 244–255, 2021.
https://doi.org/10.1007/978-3-030-92163-7_20

the main assumption that macro systems whose time evolution is dictated by the Shannon's entropy [6–8] are subjected to exhibit nonlinearities. Under this scheme it was associated the criterion of human mobility that to some extent it appears to be purely quadratic, so that it allows us to build a sustainable model that matches known statistical data of rate infections in most countries. Apart from the statistical errors, the present approach reveals us that even in the scenarios of implementation of actions that force people stay apart from each other, infections would be independent of it and it can be seen as a natural response to the human behavior that is based in mobility and randomness originated by their intrinsic behavior against other mammals and species.

The rest of paper is structured as follows: in second section the theoretical basis of this study is done. In third section the implications of model is presented having focused at the number of infections with respect to time. In fourth section the applications of model to data from Brazil, France, Mexico and Philippines is done, and finally the conclusion of paper is presented.

2 The Physics-Mathematics Machinery

Consider a finite number of local habitants defined by N and the N^ℓ the global number of people with ℓ an positive integer number. Thus one has the probability that one of them has a well-defined behavior given by:

$$\mathbf{p} - \frac{1}{N^\ell}. \tag{1}$$

Now consider that exists a finite number of habitants n^Q belonging to a social group with $Q > 0$ an integer number. Logically $n^Q \in N^\ell$. Thus, the probability that one habitant of this social group has a behavior (that for example can be matched to those of the N^ℓ is given by:

$$\mathbf{P} = \frac{1}{n^Q}. \tag{2}$$

Therefore with Eq. (1) and Eq. (2) one can built a kind of universal expression that assign a probability for each individual belonging to the

$$\mathbf{U} = \mathbf{p} \otimes \mathbf{P} = \frac{1}{n^Q N^\ell}. \tag{3}$$

Thus, it is argued that exist a Shannon's entropy [9,10] in the sense that the universal probability cannot be accurate so that instead it exhibits a well-defined uncertainty, thus the application of the Logarithm reads

$$\mathcal{H} = -\mathrm{Log}\mathbf{U} \tag{4}$$

$$= -\mathrm{Log}\left(\frac{1}{n^Q N^\ell}\right) \tag{5}$$

$$= \mathrm{Log}\left(n^Q N^\ell\right). \tag{6}$$

Thus it is feasible to postulate that exists a concrete probability of any event that affects directly the global number of habitants. As seen in Eq. (3) this probability can be a finite number of probabilities. In this way one can write down that the fraction of the affected global **H** is the outcome of single or chain or events:

$$\mathbf{H} = \left[\prod_j^J p_j \right] N^\ell. \tag{7}$$

In this way one can see that the product of probabilities is actually the product of events by the which affects the number N^ℓ due to a concrete fact that for example can be a pandemic. Under this view one can postulate that in epochs of pandemic might coexist various events that lead to establish a global pandemic. Of course, the pandemic cannot be seen as a fully negative fact, since it is proved that humans are beings capable to find the best ways to minimize the consequences of it. Then, for instance one can propose below up to three different events that might to constitute a close dynamics of pandemic:

- Mobility: Is fully inherent in all beings that carry out activities for surviving, such as bacteria that needs to carry out quorum previous to optimize any actions of chemotaxis to guarantee their motility [11, 12], humans would require of mobility to accomplish their activities as well as to take decisions. Thus the action of moving can be seen as a scenario of risk permanent to be attacked by external organisms [13–15].
- Disease: The transportation of virus and bacterias to a group of humans and the subsequent infection among them is commonly done only if the rate of mobility is high enough to produce a fast infection in a global manner.[1]
- Alleviation: In a context of pandemic, once that humans have carry out quorums to conclude in the best decisions that are favorable to them and against virus spread, humans are able to implement rules that are strongly oriented to mitigate the rate of infections and therefore to expect a notorious difference between the before and after.

2.1   Probability of Mobility

Thought as the more wide and general action in humans might not to have a very specific mathematical model. Therefore, to priori we attain an inverse relationship at the sense that it is inverse to the time to the power $\ell - s$ with s a positive real number and ℓ that is the power attained to the number of habitants (see Eq. (7) N^ℓ). In this way we postulate the probability of human mobility as

$$p_1 = \frac{1}{t^{\ell-s}}. \tag{8}$$

The why this probability falls with time is because humans is considered a specie with limitations so in most case them are depending of artifacts to perform a

---

[1] Here one can mention the aerial transportation as an important vector to increase the infection in a intercontinental way.

long distance mobility, such as planes or ships, for instance. Clearly, mobility must require external resources (fuel for example) or self-efforts (pre-history civilizations that shown the nomad characteristics in the first humans). Therefore this probability should be decaying with time.

2.2 Probability of Disease

Once that humans are drawn strategies and concrete schemes to survive as well as improve their quality of life, interactions with different families of mammals or not mammals, might to pose them near to a latent risk of being randomly infected due to micro agents such as virus or bacteria. Here, it should be noted that interactions are rather related to physical variables such as space and time. It was studied in [16] where intercontinental interactions are mainly done through flights. Thus the principle of space-time propagator based on the Feynman theory was used. In fact, while humans perform social interactions, the human-human interactions would not take a large time be large, instead it in most cases these interactions are instantaneous. In this manner it is feasible to define an universal parameter a that for larger times of interaction the interaction can be broken due to stochastic fluctuations. Subsequently, one can define below the probability of infection as

$$p_2 = \frac{1}{1 + \frac{a}{t^\ell}}, \tag{9}$$

thus for scenarios by which $a \to t^\ell$ it is obvious that p_2 the probability that would determine the infection is 0.5 that exhibits not any outcome if some is infected or not (as the one when the coin is thrown). In order to illustrate it numerically, consider two people in two different interactions of 2 s and 5 s, with ℓ a variable. Below these cases have been displayed for $a = 1000$ and $a = 100$. The left-side displays the exponential behavior of probability to be infected for large times of interaction. The right-side however displays that for 2 s. The probability is of order of 0.2 yielding a small value. Aside one can note the relevant role of parameter a (Fig. 1).

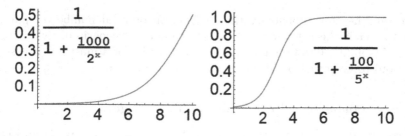

Fig. 1. Plot of Eq. (9) for the cases of $a = 1000$ (left panel) and 100 (right panel).

2.3 Probability of Mitigation

Since humans are autonomous and intelligent enough to alleviate the unexpected apparition and fast spread of strain. Only when there is common agreement in their communities or societies, unbreakable rules could be established as logic reaction to the prospective and critic scenario of an unstoppable depletion of the complete society. In this manner it is postulated that this probability would have to be favorable to humans, so that one expects the existence of a second universal parameter that would establish a high probability that guarantee the surviving of specie. On the other side, although **extinction is a remote scenario**, not any strong support is to date known that bacteria or virus have could deplete an entire specie along the first epochs of live in the planet. Based of this one can write down that the probability of mitigation can be written as:

$$p_3 = 1 - \frac{b}{t^\ell}, \tag{10}$$

that suggests that a high probability of mitigation is when $t^\ell \gg b$, that would demand a large time in the sense for example: $t = 10^6$ and $\ell = 2$ then $(t = 10^6)^{\ell=2} = 10^{12}$ although b gets the value of 10^{10} then $p_3 = 1-10^{-2} = 0.99$ yielding a high probability of mitigation.

2.4 Formulation of Complete Probability

Having defined these three probabilities, then one can construct the complete probability of event that would affect a number of habitants N^ℓ due to the changes that its society might to experience due to any virus and its propagation. Then one gets that the number of affected **H** is written as:

$$\mathbf{H} = \prod_j^3 p_j N^\ell = p_1(s,\ell)p_2(a,t,\ell)p_3(b,t,\ell)N^\ell \tag{11}$$

$$\mathbf{H} = \left(\frac{1}{t^{\ell-s}}\right)\left(\frac{1}{1+\frac{a}{t^\ell}}\right)\left(1 - \frac{b}{t^\ell}\right)N^\ell \tag{12}$$

that establishes the probability that a fraction of total number of habitants defined as **H** is the net number of affected under a scenario of risk along the events of mobility, infection and alleviation (that is not a must but it is seen as an option) to keep the specie and avoid its full depletion.

2.5 Phenomenological Determination of Integer ℓ

One can wonder whether the presence of free parameters make it an instable model. As seen previously Eq. (12) was derived inside a conjecture of entropy in which the intrinsic events associated to humans would appear as consequence of the interrelation among them. The free parameter ℓ can be for example extracted from the demography in large cities. For example of this is seen for instance at

New York city with a population of approximately 10^3 around 1737 and reaching 10^6 for 1860. Another example constitutes Paris with 10^3 and 10^6 approximately at the years of 1100 and 1840 respectively. In addition, the case of country Brazil for example its population could have been of 9000 at the year of 1500 reaching 81M at year 2019. Therefore, one has that the value $\ell = 2$ can be a potential candidate to establish a realistic scenario of global infection. The incorporation of parameters is a mechanism that have been implemented in [17–20] where it was done the incorporation of models based on Machine Learning, concepts aimed to improve processes that need of a careful analysis previous to take a critic decision. With the choice of $\ell = 2$ then Eq. (12) is rewritten as

$$\mathbf{H}(t) = N^2 \left(\frac{1}{t^{2-s}}\right) \left(\frac{1}{1+\frac{a}{t^2}}\right) \left(1 - \frac{b}{t^2}\right). \tag{13}$$

3   Implications of the Complete Probability

$\mathbf{H}(t)$ can be still rewritten in a form more understandable such as

$$\mathbf{H}(t) = N^2 t^s \left(\frac{1}{a+t^2}\right) \left(1 - \frac{b}{t^2}\right). \tag{14}$$

As defined above, the last term of product of Eq. (14) is seen as a part of an expansion in powers of the exponential function, that actually it would result in a Gaussian function. This is opportune that this exponential encloses a kind of regulation at time of the full probability $\mathbf{H}(t)$. Clearly under the assumption that $(\frac{b}{t^2})^n \sim 0$ for $n > 1$ the term $(1 - \frac{b}{t^2})$ becomes a well defined Gaussian function. This leads to rewrite again Eq. (14) in a compact probability:

$$\mathbf{H}(t) = \left(\frac{t^s}{a+t^2}\right) \mathrm{Exp}\left(-\frac{b}{t^2}\right) N^2. \tag{15}$$

Below in Fig. 2 two illustrations of Eq. (15) are displayed. One can see that b the parameter that drives the probability of mitigation appears to be relevant on the morphology of $\mathbf{H}(t)$ seen as the fraction of a global population N^2 that is affected by a random agent (such as the unexpected arrival of a virus such as the Corona virus 2019). Thus the mathematical structure of Eq. (15) in conjunction to curves of Fig. 2 can be interpreted as follows: the component $\left(\frac{N^2 t^s}{a+t^2}\right)$ is perceived as the risk of population that performs mobility in an environment of risk, causing a social emergence such as pandemic. On the other hand $\mathrm{Exp}\left(-\frac{b}{t^2}\right)$ denotes a kind of resilience as to a fast intervention of recovery once the virus is propagating inside the human group. In this way one goes to the position to define the rate of infections per unit of time due to a virus or aggressive micro organisms with a clear potential to damage systemic apparatus of a group of humans.

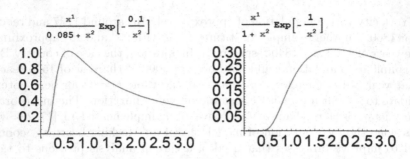

Fig. 2. Plot of Eq. (15) for the cases of $a = 0.085$, $b = 0.1$ in left panel, and $a = 1$, $b = 1$ in the right panel, here one can see the role of parameter b related to the human intervention.

3.1 Infections as Changes of Human Behavior

From the global number of habitants, then it is plausible to define the net number in risk as the derivative of $\mathbf{H}(t)$ with respect to N it is in the case that N is variable and abandons its fixed value. Then one gets:

$$\mathbf{N}(t) = \frac{d\mathbf{H}(t)}{dN} = \left(\frac{2Nt^s}{a + t^2} \right) \mathrm{Exp}\left(-\frac{b}{t^2} \right). \qquad (16)$$

Below in Fig. 3 various distributions have been displayed under the assumption that s is a variable that is attained to the mobility of humans for $b = 10$ that denotes a high intervention. It results in the flatness of all curves from a well-defined time. The flatness is an outcome of early intervention.

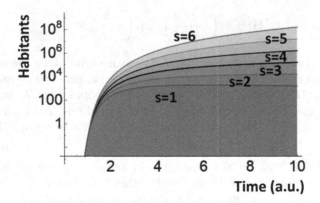

Fig. 3. A potential curve of cases by infection of virus as function of time from Eq. (16) showing the flatness of curve in certain times.

3.2   The Number of Infections

Again one can apply an additional derivative to Eq. (16) in the sense that this change would give a net number of critically affected by the presence of an external agent such as a virus causing a disease among the members of a human group. In this manner, as opted at 2020 in the beginning of ongoing Corona virus 2020 pandemic, social distancing as well as curfew actions were implemented in order to guarantee minimal human contact and therefore to expect that the curve of infections would fall down minimal values in the shortest times [21]. Thus, one expects serious changes on the infection curve in time after actions have been taken to protect people. Under this scenario, the net number of infections $\mathbf{n}(t)$ by external agent is written as:

$$\mathbf{n}(t) = \frac{d^2\mathbf{H}(t)}{dtdN} = \frac{2N}{a+t^2}\left[st^{s-1} + 2bt^{s-3} - \frac{2t^{s+1}}{a+t^2}\right]\mathrm{Exp}\left(-\frac{b}{t^2}\right). \qquad (17)$$

One can see now the apparition of up to three extra terms that are interpreted as the components that drive the morphology of $\mathbf{n}(t)$ in time and that are strongly related to the infection in particular on that terms that contain the parameter a. Thus, while attenuation is done, this does not only has a concrete effect on the mitigation of infections, it is claimed that mitigation on the social restrictions can also be seen as a disadvantageous behavior that is also implemented in the lifestyle of humans. Therefore, humans are also mitigating their selfcare due to random mobility that would exhibit a clear risk to be infected by virus. Below in Fig. 4 Eq. (17) denoting the number of infections versus time is displayed. One can see that for large values of s also large values of ℓ is needed. Thus imminently implies that N^ℓ represents the total number of a country that commonly has population at the order of various millions.

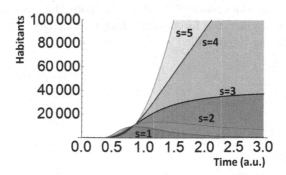

Fig. 4. Equation (17) for various values of s is displayed with s denoting the free parameter of the probability of mobility.

4   Applications and Conclusion

In Fig. 5 one can see the case of Brazil [22] whose morphology of curve of infections is to some extent similar to the case of $s = 4$. This form is seen also as a weak attenuation to create a kind of flatness of the distribution. In other words, from Eq. (17) the case of Brazil in an example where predominates the mobility and infection together as the main variables to sustain a permanent increasing number of infected people. It is supported by a small value of parameter b that poses invisible any action to mitigate the fast increasing of infections at time. Also, it is clear that from May 2020 to May 2021, the number of infections has been permanent and appears to be a line with a constant tangent.

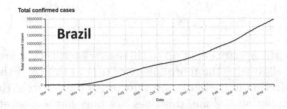

Fig. 5. Data of number of cases versus time from [22] showing the linearity as consequence of a small value of b.

In Fig. 6 the case of France is shown as sen in the data of number of infections respect to time taken from [22]. One can see that exists there up to three different periods: the first between March 2020 to August 2020 (by which is associated $s = 1$), the second one between August 2020 and December 2020 (with a potential assignation of $s = 3$), and a third one from December 2020 to February 2021 that adjusts well to the case $s = 5$. One can note that the last ones belong to scenarios that are dictated by mobility fact that makes flexible the multiple infection establishing the so-called second wave.

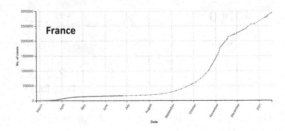

Fig. 6. The case of France fom [22] showing 3 different phases that are entirely correlated due to the evident weight of the human mobility against others variables.

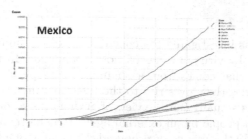

Fig. 7. The 2020 data for Mexico showing fully linear behavior for all cities that in according to Eq. (17) the present proposal indicates the high weight of mobility against other human behaviors.

In Fig. 7 the case of Mexico is shown from [22]. The data belongs to 2020 and it is given for various cities from April 2020 to September 2020, with emphasis in Mexico city that shows a fast increasing at the number of cases by the which one can adjudicate $s = 5$, denoting a little impact on the actions to mitigate the infections for instance. Again, this can also be seen as a strong magnitude of the combination of mobility and infection that is surpassing any kind of mitigation. Thus, all cities are presenting linear behaviors with minimal morphology that denotes any kind of alleviation. In Fig. 8 the case of Philippines is shown with data ranging between February 2020 to May 2021. Interestingly the data exhibits two well defined periods of infection. The first one that is dictated by $s = 1$ and a high value of b denoting a strong alleviation that produce the fall of curve in November 2020. However this alleviation appears to be weak along a **period of silent** November 2020 until February 2021. From here one can see the sharp morphology of curve that is again peaked in April 2021. One can claim that just in all those periods of silent, randomness predominates posing the system in a kind of unstable balance in the which mobility, infection and alleviation fight each other to define the identity of curve. Clearly a deep analysis of this goes beyond the scope of this paper.

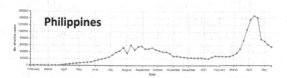

Fig. 8. The case of Philippines from [22] that exhibits two well-defined periods from the fact that the probability events might be under interference each other due to the randomness of system that does not allows to define its main identity.

4.1   Conclusion

In this paper it was investigated the contribution of the human mobility in the events of infection by virus in times of global pandemic. In this manner it was proposed a model that comes from a Shannon's entropy by the which exists a chain of probabilities that gives rise to the infection among humans. Thus the curve of infection is actually (for the view of author) the second derivative of the number of affected people by an external agent that has capabilities to deplete the specie in time. In this manner is proposed the number of infections as function of time that depends on free parameters in according to what predominates in the system: mobility, infection and alleviation. The main result of this paper Eq. (17) has been applied to the data of Brazil, France, Mexico and Philippines. It was seen that mitigation is strongly surpassed by mobility fact that sustains the infection despite of regulations such as social distance, quarantine, curfew, etc. Therefore, one can claim that human mobility is a strong factor to spread the virus in a random manner, fact that is justified by the open of terrestrial and aerial transportation at the end of first wave. Thus, people has continued to spread the virus and the different variants, fact that poses in risk the programs of vaccine. Finally the author recommends that while there is evidence of a single variant of virus, mobility would have to be blocked otherwise the apparition of more sub-types of virus can be transported among all continents making invalid the vaccines by producing again peaked distributions as seen in the case of Philippines.

References

1. Wu, F., et al.: A new coronavirus associated with human respiratory disease in China. Nature **579**, 265–269 (2020)
2. Leung, K., Wu, J.T., Liu, D., Leung, G.M.: First-wave COVID-19 transmissibility and severity in China outside Hubei after control measures, and second-wave scenario planning: a modelling impact assessment. Lancet **395**(1023325), 1382–1393 (2020)
3. Gonzalez-Parra, G., Arenas, A.J., Aranda, D.F., Segovia, L.: Modeling the epidemic waves of AH1N1/09 influenza around the world. Spat. Spatio-Temporal Epidemiol. **2**(4), 219–226 (2011)
4. Ortiz-Prado, E., et al.: Clinical, molecular, and epidemiological characterization of the SARS-CoV-2 virus and the Coronavirus Disease 2019 (COVID-19), a comprehensive literature review. Diagn. Microbiol. Infect. Dis. **98**(1), 115094 ((2020)
5. Chia, W.N., et al.: Serological differentiation between COVID-19 and SARS infections. Emerg. Microbes Infect. **9**(1), 1497–1505 (2020)
6. Jaynes, E.T.: Information theory and statistical mechanics. Phys. Rev. **106**, 620 (1957). Published 15 May 1957
7. Bekenstein, J.D.: Entropy content and information flow in systems with limited energy. Phys. Rev. D **30**, 1669 (1984). Published 1 October 1984
8. Tikochinsky, Y., Tishby, N.Z., Levine, R.D.: Alternative approach to maximum-entropy inference. Phys. Rev. A **30**, 2638 (1984). Published 1 November 1984
9. Lindgren, K.: Microscopic and macroscopic entropy. Phys. Rev. A **38**, 4794 (1988). Published 1 November 1988

10. Crutchfield, J.P., Young, K.: Inferring statistical complexity. Phys. Rev. Lett. **63**, 105 (1989). Published 10 July 1989
11. Nieto-Chaupis, H.: Macrophage-inspired nanorobots to fast recognition of bacteria and virus through electric forces and fields patterns inside of an internet of bio-nano things network. J. Phys. Conf. Ser. **1310** (2018). Applied Nanotechnology and Nanoscience International Conference (ANNIC: 22–24 October 2018. Langenbeck Virchow Haus, Berlin, Germany
12. Nieto-Chaupis, H.: The Feynman path integral to characterize and anticipate bacteria chemotaxis in a host healthy body. J. Phys. Conf. Ser. **1310** (2018). Applied Nanotechnology and Nanoscience International Conference (ANNIC: 22–24 October 2018. Langenbeck Virchow Haus, Berlin, Germany
13. Cirillo, P., Taleb, N.N.: Tail risk of contagious diseases. Nat. Phys. **16**, 606–613 (2020)
14. Morse, S.S.: The origins of new viral diseases. J. Environ. Sci. Health, Part C **9**, 2 (1991)
15. Tian, H., Xu, B.: Persistence and transmission of avian influenza A (H5N1): virus movement, risk factors and pandemic potential. Ann. GIS **21**(1), 55–68 (2015)
16. Nieto-Chaupis, H.: Feynman-theory-based algorithm for an efficient detaining of worldwide outbreak of AH1N1 virus. In: 2019 IEEE CHILEAN Conference on Electrical, Electronics Engineering, Information and Communication Technologies (CHILECON)
17. Mei, X., et al.: Artificial intelligence-enabled rapid diagnosis of patients with COVID-19. Nat. Med. **26**, 1224–1228 (2020). 19 May 2020
18. Tuli, S., Tuli, S., Tuli, R., Gill, S.S.: Predicting the growth and trend of COVID-19 pandemic using machine learning and cloud computing. IoT **11**, 100222 (2020)
19. Yadav, M., Perumal, M., Srinivas, M.: Analysis on novel coronavirus (COVID-19) using machine learning methods. Chaos, Solitons Fractals **139**, 110050 (2020)
20. Bachtiger, P., Peters, N.S., Walsh, S.L.F.: Machine learning for COVID-19–asking the right questions, The Lancet Digital Health. In press, corrected proof Available online 10 July 2020
21. Habersaat, K.B.: Ten considerations for effectively managing the COVID-19 transition. Nat. Hum. Behav. **4**, 677–687 (2020)
22. https://en.wikipedia.org wiki COVID-19 pandemic by country and territory

The Conjunction of Deterministic and Probabilistic Events in Realistic Scenarios of Outdoor Infections

Huber Nieto-Chaupis$^{(\boxtimes)}$

Universidad Autónoma del Perú, Panamericana Sur Km. 16.3 Villa el Salvador, Lima, Peru

Abstract. The aim of this paper is the derivation of an robust formalism that calculates the so-called social distancing as already determined in the ongoing Corona Virus Disease 2019 (Covid-19 in short) being established in various places in the world between 1.5 m and 2.5 m. This would constitutes a critic space of separation among people in the which aerosols might not be effective to infect healthy people. In addition to wearing masks and face protection, the social distancing appears to be critic to keep people far of infections and consequences produced from it. In this way, the paper has opted by the incorporation of a full deterministic model inside the equation of Weiss, by the which it fits well to the action of outdoor infection when wind manages the direction and displacement of aerosols in space. Thus, while a deterministic approach targets to propose a risk's probability, a probabilistic scenario established by Weiss in conjunction to the deterministic events would yield an approximated model of outdoor infection when there is a continuous source of infected aerosols that are moving through air in according to a wind velocity. The simulations have shown that the present approach is valid to some extent in the sense that only the 1D case is considered. The model can be extended with the implementation of physical variables that can attenuate the presence of disturbs and random noise that minimizes the effectiveness of present proposal.

Keywords: Covid-19 · Weiss probabilistic equation · Outdoor infection

1 Introduction

The ongoing pandemic Covid-19 has been firstly identified in Wuhan in China the last of 2019 [1,2]. The virus has been randomly propagated overall continents and phases of global infection were observed, some points that are noteworthy constitutes the number of new infections as well as the continue increasing of fatalities [3]. As well-known up to two well-defined waves have been observed: at middle of 2020, end 2020 in conjunction to the beginning of 2021. Furthermore,

© ICST Institute for Computer Sciences, Social Informatics and Telecommunications Engineering 2021
Published by Springer Nature Switzerland AG 2021. All Rights Reserved
T. Nakano (Ed.): BICT 2021, LNICST 403, pp. 256–268, 2021.
https://doi.org/10.1007/978-3-030-92163-7_21

a third wave might be happening nowadays as result of the mutation of strain. In this manner, countries are adopting social policies to face and contain the incoming worst scenarios [4]. In fact, curfew, quarantines and social restrictions were firmly adopted. Before the vaccination seasons the so-called social distancing was imposed as a response to reduce the positive exponential increasing of new cases that in most cases is consisting of vulnerable population [5] as well as for a solid return of a harmonic social-economy scenario [6]. Even the vaccination has been completed, one might to expect the arrival of new variants that would trigger again a new wave of pandemics. To counteract this, one might to appeal to distance-based policies that are demonstrated to be effective to some extent [7,8]. Some of these restrictions as:

- Vaccination of up to two doses,
- Usage of mask and face protection,
- Social distancing,
- Permanent usage of alcohol and disinfectant,
- Lock-down if needed.

The rest of this paper is as follows: In second section the deterministic scenario of outdoor infection is presented. Here is incorporated the wind velocity that emerges in an inherent manner, important equations are derived. In third section the equation of Weiss is presented and its importance as mathematical formulation that engages the action of outdoor infection is presented. In fourth section, the conjunction of deterministic and probabilistic formulations is presented in order to postulate the equation of outdoor infection. Here computational simulations are presented. The social distancing is estimated. Finally in last section the conclusion of paper is done.

2 Theory of Outdoor Infection: Main Equations

In order to construct a theory of outdoor infection, a few requirements are needed in order to derive deterministic equations. These are as follows:

- A finite number of healthy people and at least one infected,
- The infected one (or ones) are carrying out a loud speech,
- The infected one exhales a finite number of aerosols, being them that are transporting the virus,
- There is wind velocity that might to be pointing the place where are people,
- Depending at climate conditions and aerodynamics aerosols can be negligible a few meters after that were exhaled.

2.1 Derivation of Model

Consider the distribution of net number of aerosols per unit of time and distance $\mathbf{n}(r,t)$ that can be defined as:

$$\mathbf{n}(r,t) = \frac{\Delta N}{\Delta r \Delta t} \qquad (1)$$

with ΔN the number of aerosols that were exhaled. Therefore, the instantaneous number $\mathbf{N}(T,R)$ that are scattered along the space can be written as:

$$\mathbf{N}(T,R) = \int_0^T \int_0^R dtdr\mathbf{n}(r,t). \tag{2}$$

Because aerosols is a fluid that is propagating along air, then the diffusion equation can also be applied. Therefore for the unidimensional case (that of course is a special case, a 3D modeling is beyond the scope of this paper), can be written as:

$$\frac{d\mathbf{n}(r,t)}{dt} = D\frac{d^2\mathbf{n}(r,t)}{dr^2} \tag{3}$$

and from this a crude solution can be done as:

$$\mathbf{n}(r,t) = D\int dt\frac{d^2\mathbf{n}(r,t)}{dr^2} \tag{4}$$

so that Eq. 2 might be more accurate in the form as given below:

$$\mathbf{N}(R,T) = D\int_0^T dt \int_0^R dr \int_0^t d\tau \frac{d^2\mathbf{n}(r,\tau)}{dr^2}. \tag{5}$$

The integration over dt turns out to be trivial, so that one has:

$$\mathbf{N}(R,T) = DT \int_0^R dr \int_0^t d\tau \frac{d^2\mathbf{n}(r,\tau)}{dr^2}. \tag{6}$$

Nevertheless, the left-side can still be simplified in the sense that

$$\mathbf{N}(R,T) = DT \int_0^R dr \int_0^t d\tau \frac{d}{dr}\frac{d\mathbf{n}(r,\tau)}{dr} = DT \int_0^R dr \int_0^t d\tau \frac{d}{dr}\frac{d\mathbf{n}(r,\tau)}{dr} \tag{7}$$

Consider the human speech in the scenario of loud speak so that one can expect that a substantial amount aerosols are being scattered. Without physical considerations of dehydration and a certain direction of wind then one can attribute a periodical distribution for example.

2.2   Incorporation of Wind Velocity

It is trivial that a velocity appears after a straightforward insertion of the chain rule in Eq. 2:

$$\mathbf{N} = \int_0^T dt \int_0^R dr \frac{d\mathbf{n}}{dr}\frac{dr}{dt} = \int_0^T dt \int_0^R drv(t)\frac{d\mathbf{n}}{dr} = \int_0^T dt \frac{d\mathbf{n}}{dr}v(t). \tag{8}$$

In this manner from Eq. 6 one can generalize to a finite number of sources, so that one gets:

$$\mathbf{N}(r) = \sum_q \int_0^T dt \frac{d\mathbf{n}_q(r,t)}{dr}v_q(t), \tag{9}$$

where q runs over the possible sources of aerosols [9] and that in praxis denote all those people that were infected and contain the strain and actions of loud speech would be an imminent manner to infect healthy people. One can wonder about the change in time of $\mathbf{N}(r)$ that can be written as

$$\frac{d\mathbf{N}(r)}{dt} = \sum_q \frac{dn_q(r,t)}{dr} v_q(t), \qquad (10)$$

with $v(t)$ the 1-dimension direction of wind. This leads to define a new relation that involve a fraction of net number of aerosols that might to be under the influence of wind:

$$\frac{dn_q}{dr} = \frac{\eta \mathbf{N}}{\tilde{n}_q} \qquad (11)$$

so that one can see that under the presence of wind, $\mathbf{n}$ is now a fraction of all those sources exhaling aerosols with $\tilde{n}_q$, and η a random number. Thus Eq. 11 in Eq. 12 one arrives to:

$$\frac{d\mathbf{N}(r,t)}{dt} = \sum_q \frac{\eta \mathbf{N}(r,t)}{n_q} v_q(t) = \eta \mathbf{N}(r,t) \sum_q \frac{v_q(t)}{n_q}. \qquad (12)$$

This indicates that a change at $\mathbf{N}(r,t)$ is proportional to it times the wind velocity $v_q(t)$. Therefore one can go through a trivial integration in both sides:

$$\int \frac{d\mathbf{N}(r,t)}{\mathbf{N}(r,t)} = \int_0^t \eta \sum_q \frac{v_q(t)}{n_q} dt, \qquad (13)$$

yielding the familiar exponential distribution:

$$\mathbf{N}(r,t) = N_0 \mathrm{Exp}\left[\frac{\pm\eta}{Q} \sum_q^Q \int_0^{\mathbf{T_{DH}}} \frac{v_q(t)}{n_q} dt \right] \qquad (14)$$

where the integer Q at the denominator indicates the average over all sources. Here $\mathbf{T_{DH}}$ denotes the time of dehydration for all aerosols.

In Fig. 1 the logarithm of Eq. 14 has been plotted under the assumption that $Q = 2, 3, 4$ and 5. The velocity $v_q = 0.05(t + 2 * q)^2 \mathrm{Sin}\,[q * t]$ and $n_q = 1 + 20^q$ the number of aerosols. Also, $\pm\eta/Q = -1/2$, $N_0 = 1/1.2$. One can see that the distribution exhibits a kind of minimum at 3.5 s.

One can attribute these minimum as the time by the which the dehydration takes place. The why one can see the up of distribution from $t = 4$ s is because the wind velocity v_q that has been modeled through a sinusoidal function.

And it is actually encompassing the fact that in outdoor the wind direction is a composition of various ones due to the complexity of terrain and obstacles.

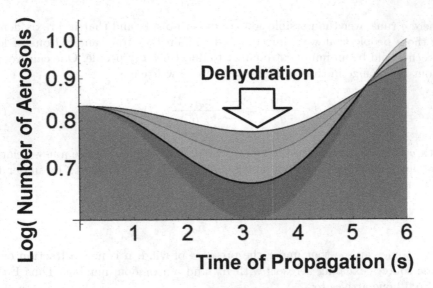

Fig. 1. Illustration of Eq. 14 showing a minimal at the normalized logarithm distribution of number of aerosols for $t = 3.5$ s supposed to be caused by the dehydration of aerosols in air. The up seen at the distributions can be explained in terms of wind velocity. Plot and the next ones were done with Wolfram [11]

3   The Stochastic View of Weiss Equation

As it is well-known, outdoor infection can be done with a high effectiveness in such scenarios as displayed in Fig. 2. In fact, while there is fix wind velocity, aerosols can be displaced some meters before their dehydration producing massive infection.

3.1   The Weiss Equation

In circumstances as shown in Fig. 2, to talk about that outdoor infection is done through deterministic laws might have a weak sustain so that, one can complement to it a formulation created under probabilistic grounds. In fact, the equation of Weiss considers the following action: Assume a finite number of N objects are thrown to a finite number of Q cells. All cells exhibit same morphology so that no any preference exists there. When this action is repeated M times, then one arrives to the probability **P** that a cell is unoccupied[1]. This can be written below as:

$$\mathbf{P} = \frac{Q\mathrm{Log}Q}{NM} \tag{15}$$

---

[1] Obviously the probability of occupancy is done by $1-\mathbf{P}$.

Fig. 2. Example of a potential scenario of random infection when at least one participant of chorus is spreading aerosols transporting virus. Taken from [10].

3.2 Intuitive Derivation

Consider the simplest case: the action that N objects are thrown to Q cells. One can wonder how many tries can be done if one requires that all cells are occupied. Clearly for this it is demanded a large number of M tries. Then the probability that exists unoccupied ones is given by:

$$\mathbf{P} = \frac{Q}{NM}. \tag{16}$$

For example, consider $Q = 10$ cells and $N = 1000$ objects. Thus for the first event one has that $\mathbf{P} = 10^{-2} = 0.01$ the probability of having 10 unoccupied cells. When Eq. 16 is written as:

$$\int d\mathbf{P} = \int \frac{dQ}{NM}. \tag{17}$$

Thus, both M and N are totally independent of Q, so that one can write below:

$$\frac{1}{Q}\frac{d\mathbf{P}}{dQ} = \frac{1}{NM}\frac{1}{Q} \tag{18}$$

by the which both members were divided by Q. With this one has that:

$$\frac{d\mathbf{P}}{Q} = \frac{1}{NM}\frac{dQ}{Q} \tag{19}$$

whose integration in both members yields Eq. 15. On the other hand the probability there is Q occupied cells can be written as $\mathbf{p} + \mathbf{P} = 1$, so that from this the one gets:

$$\mathbf{p} = 1 - \frac{Q \mathrm{Log} Q}{NM}. \tag{20}$$

Interestingly, from the following assumption:

$$\left(\frac{QLogQ}{NM}\right)^{\ell} \approx 0 \tag{21}$$

$\forall \ell \geq 2$ that is true only if $NM >> QLogQ$. With this then one can arrive to a new definition for the occupancy probability:

$$\mathbf{p} = \mathrm{Exp}\left(-\frac{QLogQ}{NM}\right). \tag{22}$$

In Fig. 3, the occupied, unoccupied and total probabilities are plotted. All of them are function of number of cells. From it one can see that all have a coincidence for $Q = 118$ cells with a common probability of 0.56. For this exercise $NM = 1000$, equivalent to $N = 200$ objects and $M = 5$ times that were thrown the objects to the available cells. It is easy to see that occupancy decreases with the number of cells. At the other side, the unoccupancy increases with Q resulting to be rather logic since for example, the probability of having 45 unoccupied of 100 total cells is 0.45 when have been thrown 200 objects 5 times. In the inverse case one gets 65 occupies cells of 100 available ones[2].

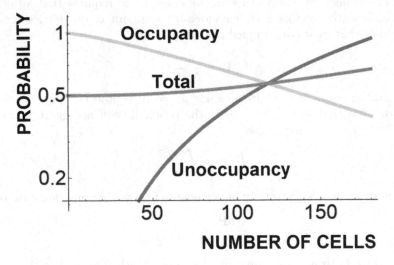

Fig. 3. The Weiss probabilities (i) Occupancy (green), (ii) Unoccupancy (magenta) and (iii) Total (orange), for $N = 200$ thrown objects in up to 5 times (Color figure online)

---

[2] One can see that sum yields 110 cells, fact that can be understood in terms of errors of model.

4   The Stochastic and Deterministic Synergy

The Weiss equation Eq. 15 has a notable application[3] in the field of Infection Dynamics, a discipline that is nowadays in constant progress with the apparition of Covid-19 pandemic. With the emergence of global pandemic, a series of policies were imposed. Among them one can see that well-known social distancing that has been established of being between 1.5 m to 2 m. Once known that virus is mainly transmitted by aerial paths because the exhalation and the subsequent emission of aerosols, then one clearly has two scenarios: (i) indoor: that has been deeply studied, and (ii) outdoor: whose scenario is not well established since in most studied the wind velocity and dehydration time were not taken into account. In this manner one can associate the Weiss equation to the following scenario: one or various infected people that are carrying out loud speech. This action is generating the production of aerosols that do not only stay there but also can be traveling along the wind direction if this is enough stable to produce the displacement of virus enclosed in the aerosol. From Eq. 15 one can establish that N objects are denoting the N aerosols, Q cells denote Q healthy people, M denotes the times that aerosols is arriving to people, and finally $\mathbf{P}$ the probability that people is not reached by infected aerosol.

4.1   The Frequency of Aerosols Arrival

It should be noted that in Eq. 15, the integer number M denoting the number of times by the which the aerosols are falling into the healthy people can also be seen as a frequency. In fact, because mathematically one does not find not any contradiction neither inconsistency, then it is feasible to rewrite again Eq. 15 as with $Q \rightarrow y$.

$$\mathbf{P}(\omega) = \frac{\omega q \text{Log}(q)}{N} \qquad (23)$$

that can be read as the probability $\mathbf{P}(\omega)$ of q healthy people becomes infected after of N aerosols have fallen on them. Here one can see that the probability is now depending on the frequency in the which the aerosols are arriving to people. Imminently while ω increases the probability does it. Taking into account all this, one can adjudicate to ω the name of **risk frequency**. Because N the number of aerosols has been proposed in Eq. 14, then it enters inside the denominator of Eq. 23:

$$\mathbf{P}(\omega) = \frac{\omega q \text{Log}(q)}{N_0 \text{Exp}\left[\frac{\pm \eta}{Q} \sum_q^Q \int_0^{\mathbf{T_{DH}}} \frac{v_q(t)}{\mathbf{n}_q} dt\right]}. \qquad (24)$$

The sign $\pm$ is reverted when Eq. 24 acquires the following form:

$$\mathbf{P}(\omega) = \frac{\omega q \text{Log}(q) \text{Exp}\left[\frac{\pm \eta}{Q} \sum_q^Q \int_0^{\mathbf{T_{DH}}} \frac{v_q(t)}{\mathbf{n}_q} dt\right]}{N_0}. \qquad (25)$$

---

[3] Among others not less relevant, such as nano particles and drug delivery carriers.

The probability for all allowed frequencies after the integration over them can be written as:

$$\mathbf{P} = \int \mathbf{P}(\omega)d\omega = \int \frac{\omega q \mathrm{Log}(q)\mathrm{Exp}\left[\frac{\pm\eta}{Q}\sum_q^Q \int_0^{\mathrm{TDH}} \frac{v_q(t)}{\mathbf{n}_q}dt\right]}{N_0}d\omega$$

$$= \int \mathbf{P}(\omega)d\omega = \frac{\omega^2 q \mathrm{Log}(q)\mathrm{Exp}\left[\frac{\mp\eta}{K}\sum_k^K \int_0^{\mathrm{TDH}} \frac{v_k(t)}{\mathbf{n}_k}dt\right]}{N_0}. \tag{26}$$

In this manner the probability can be written as:

$$\mathbf{P} = \frac{\omega^2 q \mathrm{Log}(q)}{N_0}\mathrm{Exp}\left[\frac{(\mp\eta)}{K}\sum_k^K \int_0^{\mathrm{TDH}} \frac{v_k(t)}{\mathbf{n}_k}dt\right]. \tag{27}$$

In order ro test the model in a quantitative manner, only one source is considered, thus one gets:

$$\mathbf{P} = \frac{\omega^2 q \mathrm{Log}(q)}{N_0}\mathrm{Exp}\left[(\mp\eta)\int_0^{\mathrm{TDH}} \frac{v(t)}{\mathbf{n}}dt\right]. \tag{28}$$

Here it is important to recall that the left-side admits a first time derivative so that one gets:

$$\frac{d}{dt}\mathbf{P} = \frac{d}{dt}\left(\frac{\omega^2 q \mathrm{Log}(q)}{N_0}\mathrm{Exp}\left[(\mp\eta)\int_0^{\mathrm{TDH}} \frac{v(t)}{\mathbf{n}}dt\right]\right) \tag{29}$$

because $\mathbf{P}$ depends explicitly on r the one gets an extra velocity at the left-side due to chain rule:

$$\frac{d\mathbf{P}}{dr}\frac{dr}{dt} = \frac{\omega^2 q \mathrm{Log}(q)}{N_0}\frac{d}{dt}\left(\mathrm{Exp}\left[(\mp\eta)\int_0^{\mathrm{TDH}} \frac{v(t)}{\mathbf{n}}dt\right]\right). \tag{30}$$

The velocity at the left-side can be recognized as the wind velocity:

$$\frac{d\mathbf{P}}{dr}\mathbf{V} = \frac{\omega^2 q \mathrm{Log}(q)}{N_0}\left(\mp\eta\frac{v(t)}{\mathbf{n}}\right)\left(\mathrm{Exp}\left[(\mp\eta)\int_0^{\mathrm{TDH}} \frac{v(t)}{\mathbf{n}}dt\right]\right). \tag{31}$$

With this, an universal hybrid formulation for the outdoor infection has been obtained. One can write again Eq. 31 as:

$$\frac{d\mathbf{P}}{dr} = \mp\eta\frac{\omega^2 q \mathrm{Log}(q)}{N_0\mathbf{n}}\left(\frac{v(t)}{\mathbf{V}}\right)\left(\mathrm{Exp}\left[(\mp\eta)\int_0^{\mathrm{TDH}} \frac{v(t)}{\mathbf{n}}dt\right]\right). \tag{32}$$

One can note that now the derivative with respect to distance r is directly proportional to the rate $v/\mathbf{V}$ with $\mathbf{V}$ the arriving velocity of aerosols. It is actually the final velocity of how are arriving aerosols to people, therefore it is also denoting the velocity of all those aerosols once were exhaled.

Under some circumstances one can assume that:

$$(\mp\eta)\int_0^{\mathbf{T_{DH}}} \frac{v(t)}{\mathbf{n}}dt = (\mp\eta)\int_0^{\mathbf{T_{DH}}} \frac{dr}{\mathbf{n}dt}dt = (\mp\eta)\int_0^{r_{DH}} \frac{dr}{\mathbf{n}} = \frac{\mp\eta \mathbf{r_{DH}}}{\mathbf{n}}. \quad (33)$$

Inserting Eq. 33 into Eq. 32, one gets:

$$\frac{d\mathbf{P}}{dr} = \mp\eta\frac{\omega^2 q \mathrm{Log}(q)}{N_0 \mathbf{n}}\left(\frac{v(t)}{\mathbf{V}}\right)\left(\mathrm{Exp}\left[\frac{\mp\eta \mathbf{r_{DH}}}{\mathbf{n}}\right]\right). \quad (34)$$

As done in Eq. 1 $\mathbf{n}$ has units of number of aerosols per units of time and distance. In this manner, one can wonder from Eq. 34 if $\frac{d\mathbf{P}}{dr}$ exhibits a particular morphology when it is dependent on $\mathbf{n}$. With Eq. 34 the full probability of risk (with the potential of being infected in outdoor spaces) can be expressed as:

$$\mathbf{P} = \int\left[\mp\eta\frac{\omega^2 q \mathrm{Log}(q)}{N_0 \mathbf{n}}\left(\frac{v(t)}{\mathbf{V}}\right)\left(\mathrm{Exp}\left[\frac{\mp\eta \mathbf{r_{DH}}}{\mathbf{n}}\right]\right)\right]dr, \quad (35)$$

in the which one arrives to the infection probability in the outdoor case:

$$\mathbf{P_{INF}} = 1 - \int\left[\mp\eta\frac{\omega^2 q \mathrm{Log}(q)}{N_0 \mathbf{n}}\left(\frac{v(t)}{\mathbf{V}}\right)\left(\mathrm{Exp}\left[\frac{\mp\eta \mathbf{r_{DH}}}{\mathbf{n}}\right]\right)\right]dr. \quad (36)$$

5   Simulations and Discussion

Equation 34 denotes the change of probability of q people are not being infected due to $\mathbf{n}$ aerosols [12]. Thus, while is kept both as independent variables, in Fig. 4 the contour plot is showing the distribution of the probabilities. Here, not any particular direction of wind was assumed [13]. Therefore, one can see the lowest probabilities along the blue color. Nevertheless the red color at the values of $\mathbf{n} = 1$ and $q = 50$ is telling us that $d\mathbf{P}/dr$ has a maximum so that a low number of aerosols can be reaching all 50 people. It is the case when $v(t)/\mathbf{V}$ exhibits same magnitude [14].

In Fig. 5 a density plot of Eq. 34 is displayed. Here have been considered as independent variables both $\mathbf{r_{DH}}$ and $v(t)$ the dehydration distance and wind velocity respectively. What one can learn from it is the existence of up to three different phases categorized in three colors: blue, orange and red. Clearly a probability of 0.5 for the range between 1.5 m and 2.0 m is seen as a potential scenario. In Fig. 6 same as previous but done as a contour plot. Here the apparition of various colors helps us to differentiate the sectors of high risk as for example let us to case of a wind velocity of 1 m/s. For this then would correspond a dehydration length of 1.5 m with a low probability of not being infected. It should be noted that even for small wind velocities and short dehydration length there is high probability of not being infected. These numerical results ar fully in according with policies of the so-called **social distancing** the which establishes distances of 1.5 m^4.

---

[4] However neither temperatures nor specifying indoor and outdoor were done.

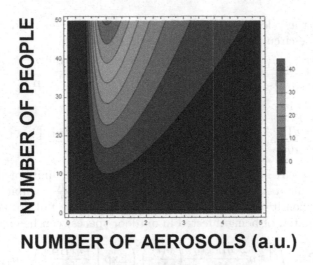

Fig. 4. The Weiss probabilities for $N = 200$ thrown objects in up to 5 times as function of number of people under risk of infection and number of aerosol (in arbitrary units). As seen the red color with a low number of aerosols, still the risk is high. (Color figure online)

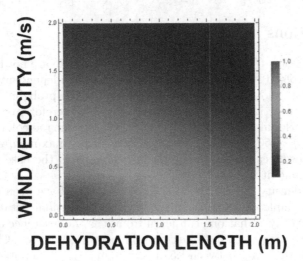

Fig. 5. The Weiss probabilities as function of wind velocity and dehydration length. One can see that in the lowest wind velocities and short dehydration lengths the risk is negligible (in full concordance to common sense and realistic cases) for wind velocities greater than 1.4 m/s fact that would have influence on the dehydration time.

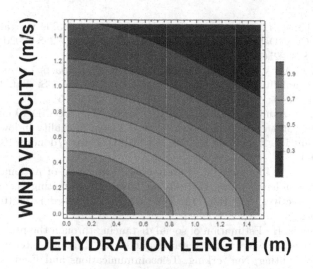

Fig. 6. The Weiss probabilities for $N = 200$ thrown objects in up to 5 times, same as Fig. 5 but done in contour plot showing the bands that belong to a concrete risk of being infected. The blue area is perceived as (Color figure online)

6   Conclusion

In this paper, a study about the estimation of the so-called social distancing was presented. In essence it consider the equation of Weiss as principal structure of this investigation. In fact, it was found that there is a firm correspondence between the Weiss equation and the phenomenon of outdoor infection that although suggested as a policy of protection against the human-human transmission of Covid-19, it is not clear if the established distances are considering the wind velocity that in some scenarios that might be realistic in praxis, can boost massive infections in open spaces. Thus, from the results of this paper is suggested to implement well-designed policies in outdoor spaces in order to guarantee that in crowded places the probability of infection is negligible in comparison to the indoor scenario [4], even with the random presence of infected people, and a strong wind velocity as well. In a future work the problem of outdoor infection but in a 3D scenario shall be treated.

References

1. Xu, B., Gutierrez, B., Mekaru, S., et al.: Epidemiological data from the COVID-19 outbreak, real-time case information. Sci. Data **7**, 106 (2020). https://doi.org/10.1038/s41597-020-0448-0
2. Jia, J.S., Lu, X., Yuan, Y., et al.: Population flow drives spatio-temporal distribution of COVID-19 in China. Nature **582**, 389–394 (2020). https://doi.org/10.1038/s41586-020-2284-y

3. Yan, L., Zhang, H.T., Goncalves, J., et al.: An interpretable mortality prediction model for COVID-19 patients. Nat. Mach. Intell. **2**, 283–288 (2020). https://doi.org/10.1038/s42256-020-0180-7

4. Wellenius, G.A., Vispute, S., Espinosa, V., et al.: Impacts of social distancing policies on mobility and COVID-19 case growth in the US. Nat. Commun. **12**, 3118 (2021). https://doi.org/10.1038/s41467-021-23404-5

5. Osendarp, S., Akuoku, J.K., Black, R.E., et al.: The COVID-19 crisis will exacerbate maternal and child undernutrition and child mortality in low- and middle-income countries. Nat. Food **2**, 476–484 (2021). https://doi.org/10.1038/s43016-021-00319-4

6. Kochańczyk, M., Lipniacki, T.: Pareto-based evaluation of national responses to COVID-19 pandemic shows that saving lives and protecting economy are non-trade-off objectives. Sci. Rep. **11**, 2425 (2021). https://doi.org/10.1038/s41598-021-81869-2

7. Nieto-Chaupis, H.: Estimation of social distancing through the probabilistic Weiss equation: it is the wind velocity a relevant factor? In: 2020 International Conference on Computing, Networking, Telecommunications and Engineering Sciences Applications (CoNTESA) (2020)

8. Nieto-Chaupis, H.: Theory of virus public infection through the Weiss approach. In: 2020 IEEE 20th International Conference on Bioinformatics and Bioengineering (BIBE) (2020)

9. Crawford, C., Vanoli, E., Decorde, B., et al.: Modeling of aerosol transmission of airborne pathogens in ICU rooms of COVID-19 patients with acute respiratory failure. Sci. Rep. **11**, 11778 (2021). https://doi.org/10.1038/s41598-021-91265-5

10. German choirs go quiet as singing branded virus risk Source: AFP Published: 2020/5/28 17:33:40. Members of the Staats- und Domchor sing "Ascension Day" on the steps in front of the Berlin Cathedral after a service on May 21. Photo: AFP. https://www.globaltimes.cn/content/1189811.shtml

11. https://www.wolfram.com/mathematica/

12. Zhang, R., Li, Y., Zhang, A.L., Wang, Y., Molina, M.J.: Identifying airborne transmission as the dominant route for the spread of COVID-19. Proc. Nat. Acad. Sci. **117**(26), 14857–14863 (2020). https://doi.org/10.1073/pnas.2009637117

13. Bourouiba, L.: The fluid dynamics of disease transmission. Ann. Rev. Fluid Mech. **53**, 473–508 (2021). https://doi.org/10.1146/annurev-fluid-060220-113712

14. Grosskopf, K.R., Herstein, K.R.: The aerodynamic behavior of respiratory aerosols within a general patient room. Science **18**, 709–722 (2012). https://doi.org/10.1080/10789669.2011.587586

Author Index

Printed in the United States
by Baker & Taylor Publisher Services